PICTURE
Book #1 MATH
Puzzles for Adults

an activity book for the Christmas holidays

J.J. Wiggins

CONTENTS

LEVEL 1

Each puzzle, you'll be given three equations with symbols and numbers. Use everything you know to figure out the final answer in the fourth equation. To do this, remember these five important rules...

1. Every symbol has a value of 1 or higher.

2. Two different symbols can have the same value.

3. There are no fractions or negative numbers.

4. Subtracting a larger number from a smaller number is not allowed.

5. The final answer is always 0 or higher.

Alright, let's get started!

#1

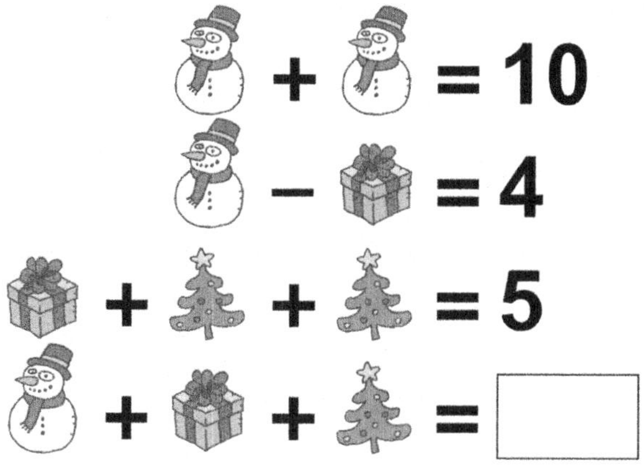

#2

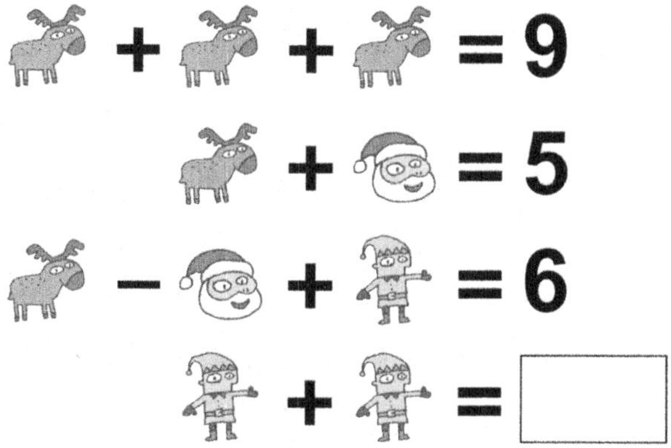

#3

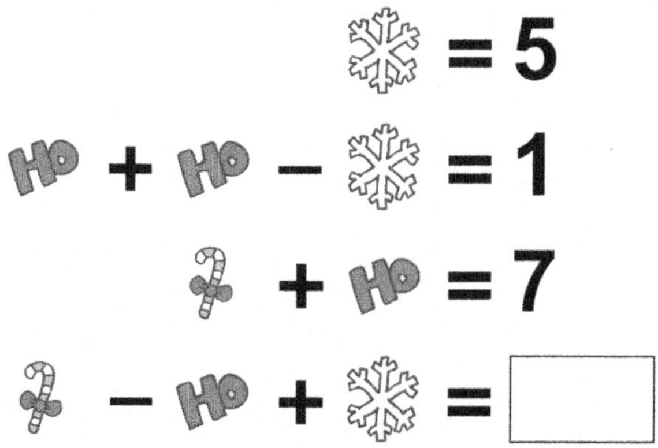

5

#4

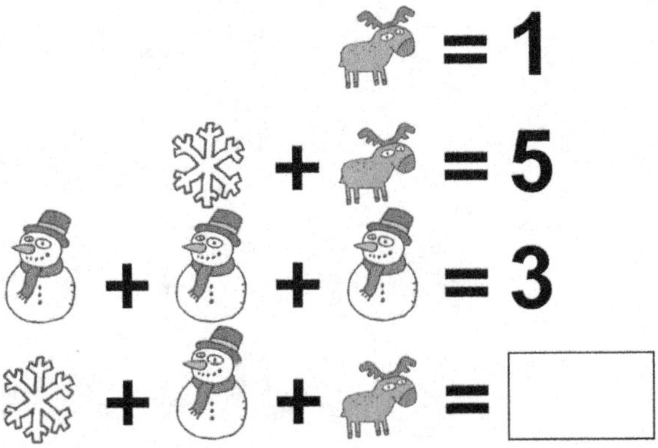

🦌 = 1

❄️ + 🦌 = 5

⛄ + ⛄ + ⛄ = 3

❄️ + ⛄ + 🦌 = ☐

#5

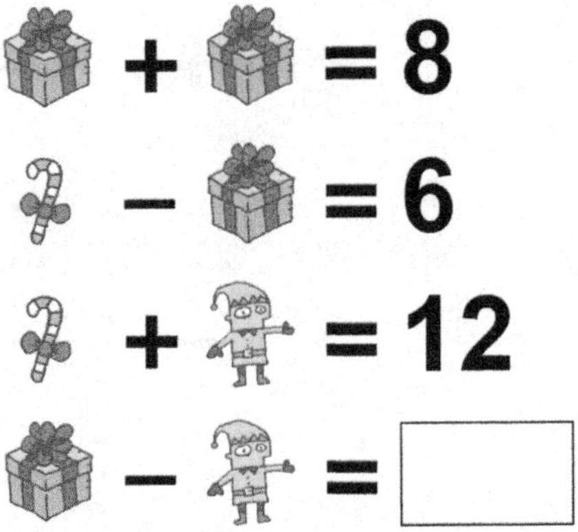

🎁 + 🎁 = 8

🍬 − 🎁 = 6

🍬 + 🧝 = 12

🎁 − 🧝 = ☐

#6

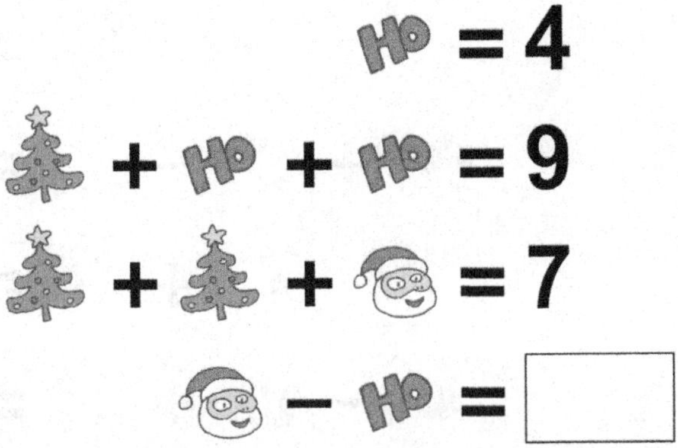

HO = 4

🎄 + HO + HO = 9

🎄 + 🎄 + 🎅 = 7

🎅 − HO = ☐

#7

 = 4

 + − = 4

 − + = 4

 + + = ☐

#8

 + + = 3

 − = 9

 − + = 8

 +  = ☐

#9

 = 2

 − = 7

 − = 5

 +  +  = ☐

#10

 + = 5

 − = 3

 − = 1

 − = ☐

#11

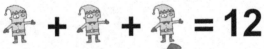 − = 2

 + + = 12

 − = 1

 + = ☐

#12

 +  =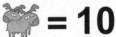

 = 10

 + = 3

 + + = ☐

#13

$$\text{HoHo} = 6$$

$$\text{❄} - \text{Ho} = 1$$

$$\text{Ho} + \text{❄} - \text{🍬} = 5$$

$$\text{🍬} + \text{🍬} + \text{🍬} = \boxed{}$$

#14

$$\text{⛄} + \text{⛄} + \text{⛄} = 15$$

$$\text{🎁} + \text{🎁} + \text{🎁} = 6$$

$$\text{🎄} + \text{🎄} + \text{🎄} = 12$$

$$\text{⛄} - \text{🎁} - \text{🎄} = \boxed{}$$

#15

$$\text{🎅} + \text{🎅} + \text{🎅} = 9$$

$$\text{🎅} - \text{🎅} + \text{🦌} = 5$$

$$\text{🦌} + \text{🧝} = 13$$

$$\text{🧝} - \text{🎅} = \boxed{}$$

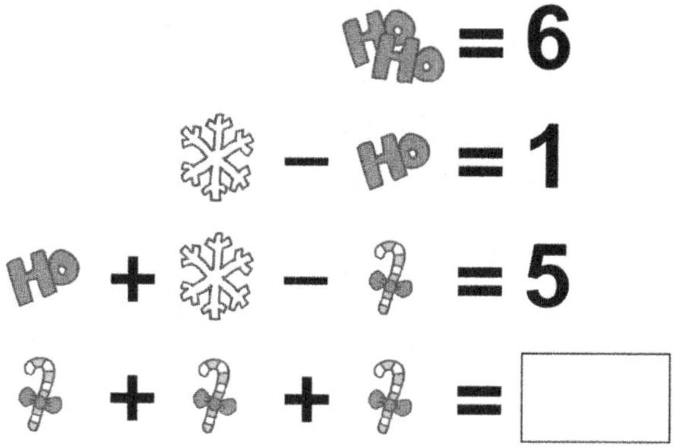

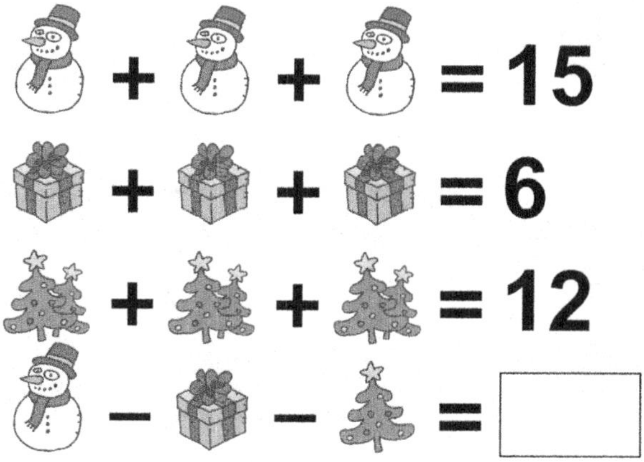

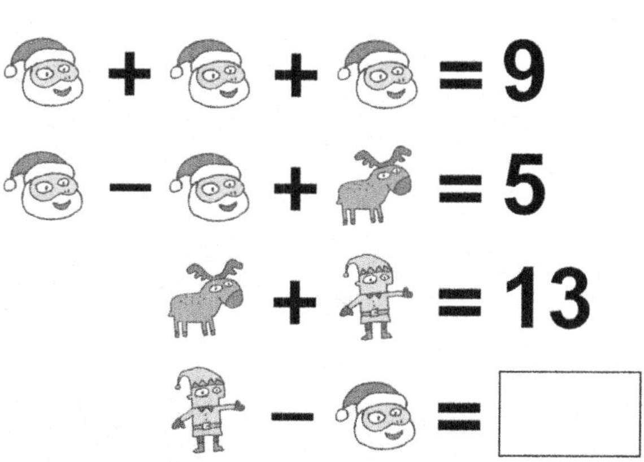

#16

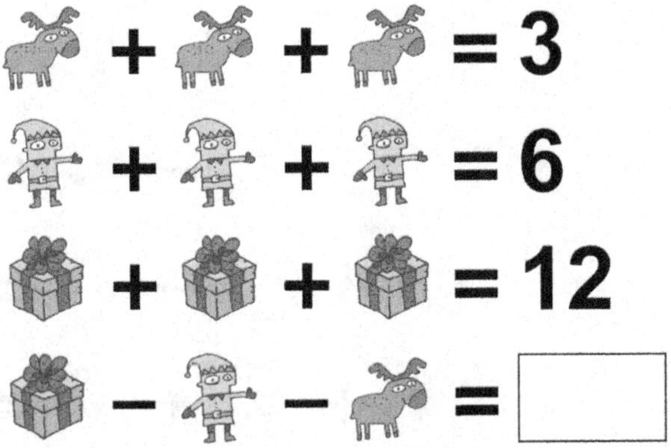

#17

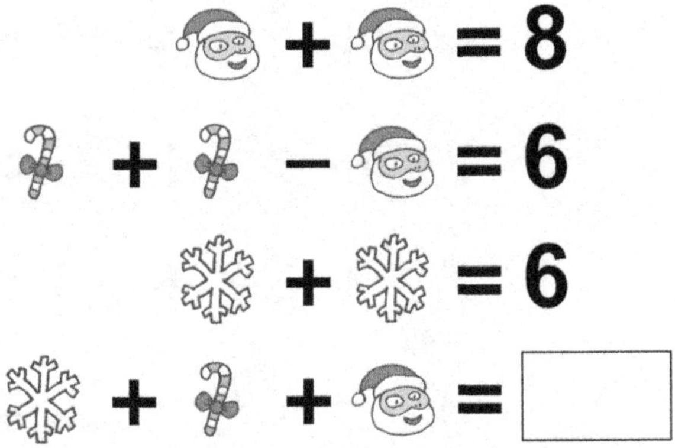

#18

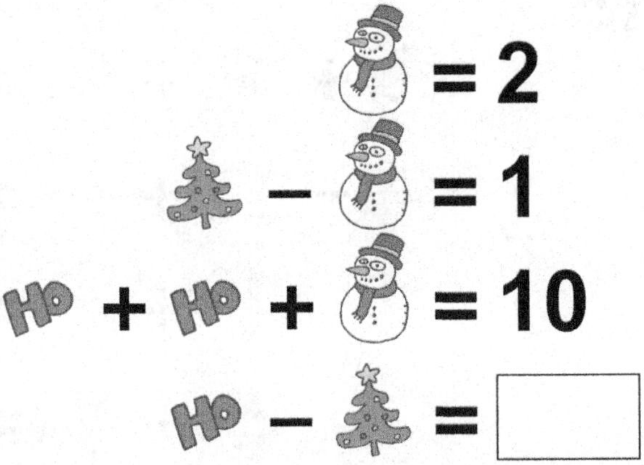

10

#19

$$\text{Ho} - \text{gift} = 1$$

$$\text{Ho} + \text{Ho} = 6$$

$$\text{gift} + \text{santa} = 7$$

$$\text{santa} + \text{santa} + \text{santa} = \boxed{}$$

#20

$$\text{candy} + \text{elf} = 8$$

$$\text{elf} + \text{elf} = 6$$

$$\text{candy} - \text{snowflake} = 2$$

$$\text{elf} + \text{snowflake} = \boxed{}$$

#21

$$\text{tree} = 6$$

$$\text{tree} + \text{moose} + \text{moose} = 11$$

$$\text{snowman} - \text{moose} = 13$$

$$\text{snowman} + \text{snowman} = \boxed{}$$

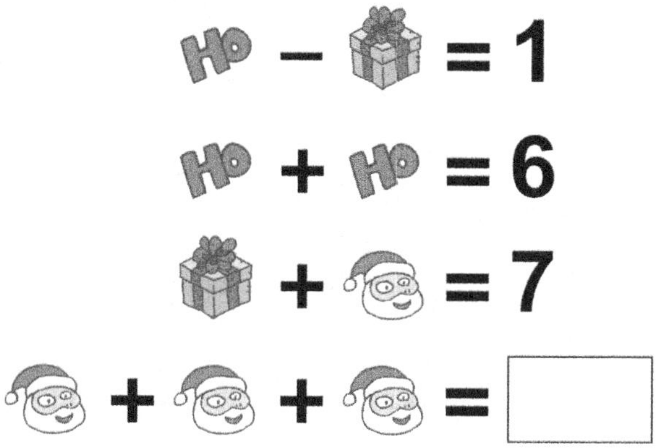

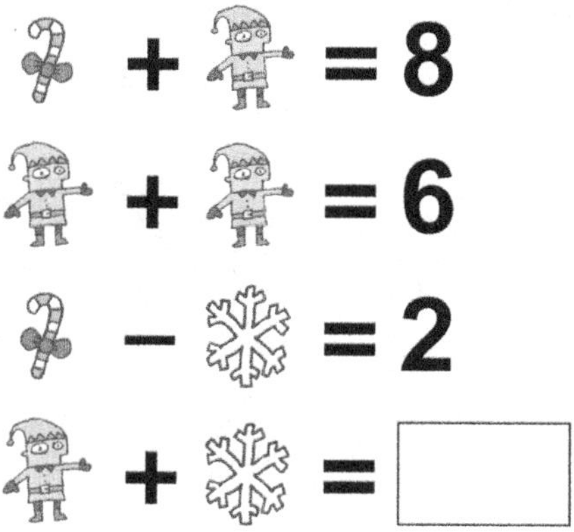

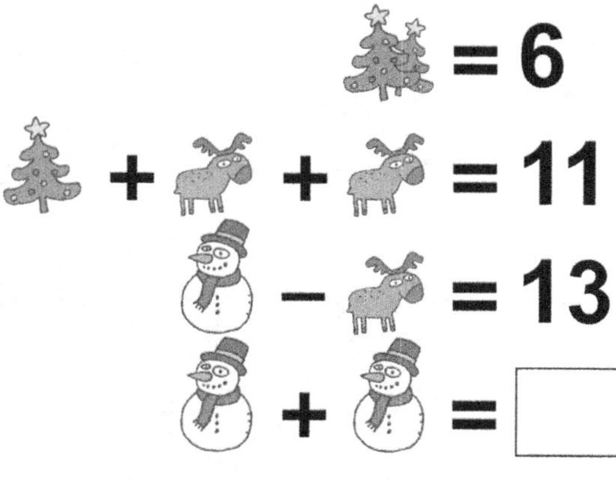

#22

🍬 = 10

🍬 + 🎁 = 15

🍬 + 🎁 + ⛄ = 16

⛄ + ⛄ + ⛄ = ☐

#23

🎄 − 🎅 = 1

🎄 + 🎄 = 6

🧝 + 🎄 = 8

🧝 − 🎅 = ☐

#24

🦌 = 10

HO + HO + 🦌 = 14

HO + HO + ❄ = 8

🦌 + ❄ = ☐

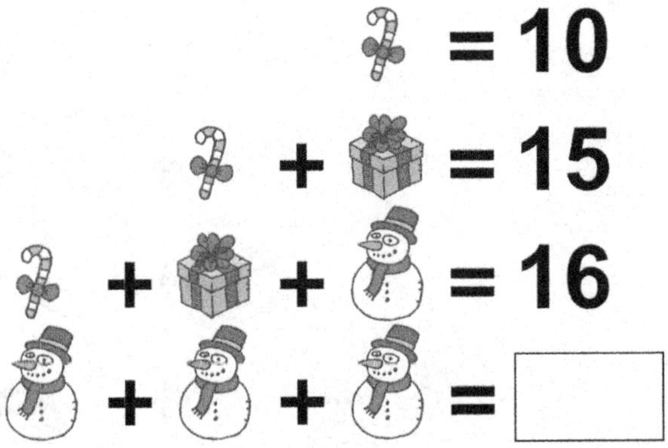

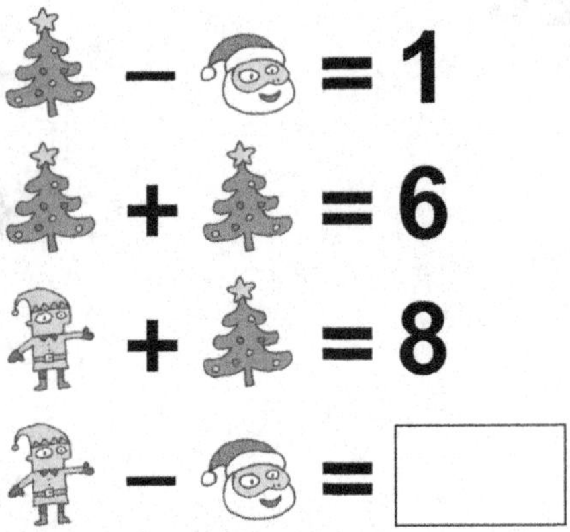

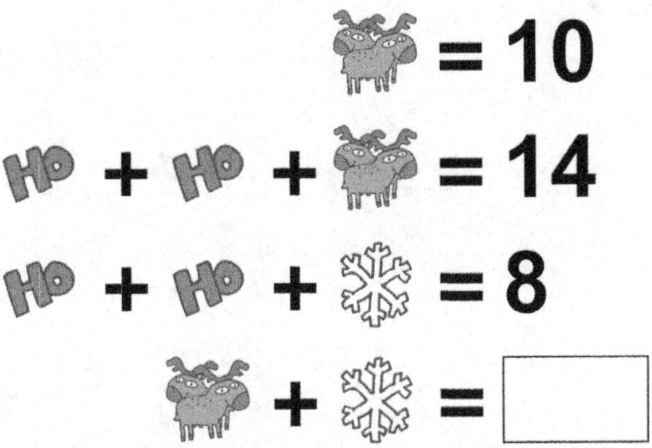

#25

🐐 − 🦌 = 3

🎅 + 🦌 + 🎅 = 5

🎄 + 🎄 + 🎅 = 11

🎄 − 🦌 = ☐

#26

HO − ❄️ = 5

❄️ − 🧝 = 1

🧝 + 🧝 = 4

HO + HO + HO = ☐

#27

🍬 + 🍬 + 🍬 = 21

🍬 − 🎁 + ⛄ = 4

🎁 + 🎁 + 🎁 = 15

🍬 − ⛄ − ⛄ = ☐

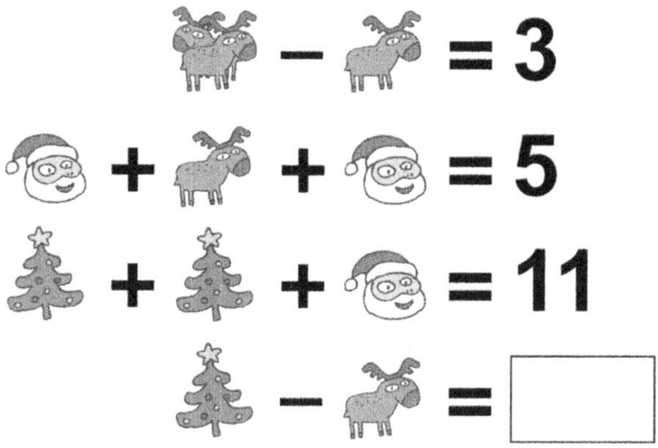

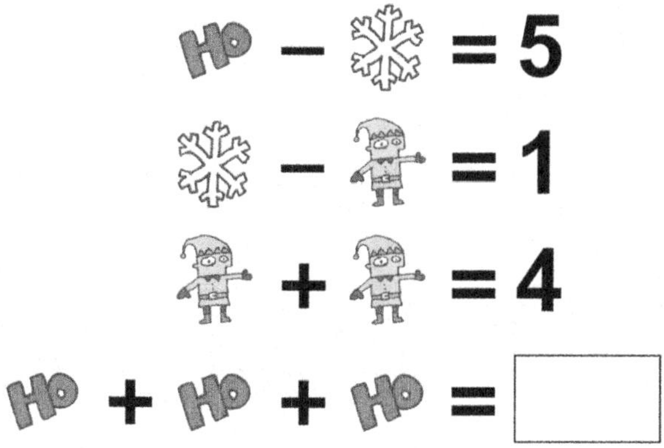

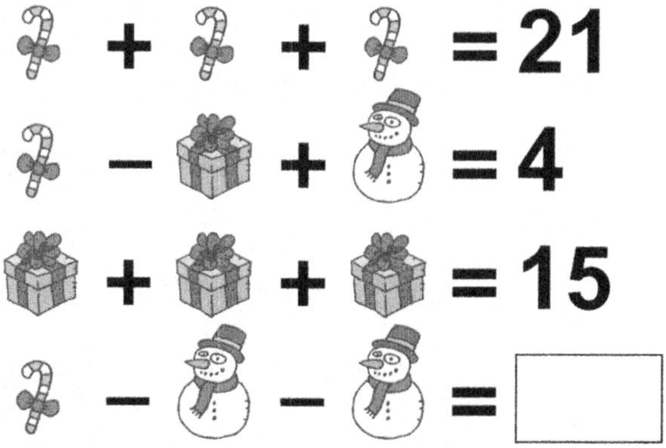

#28

(elf) + (elf) = 8

(candy cane) + (candy cane) = 14

HO − (elf) − (candy cane) = 4

HO + HO + HO = ☐

#29

(gift) + (snowflake) = 5

(snowflake) = 6

(snowman) + (gift) = 7

(snowflake) − (snowman) = ☐

#30

(reindeer) + (santa) = 3

(tree) − (reindeer) = 1

(tree) − (santa) = 2

(tree) = ☐

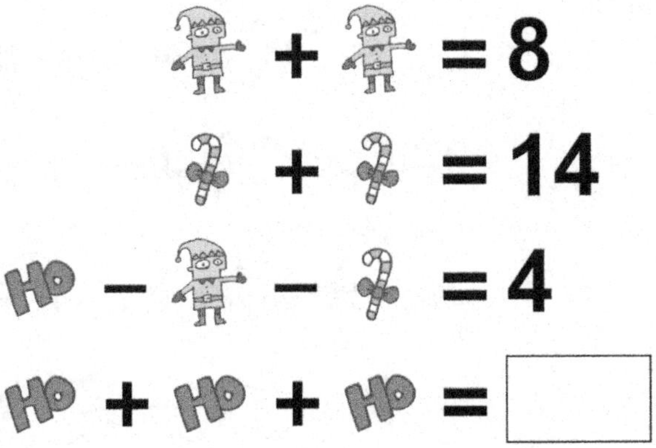

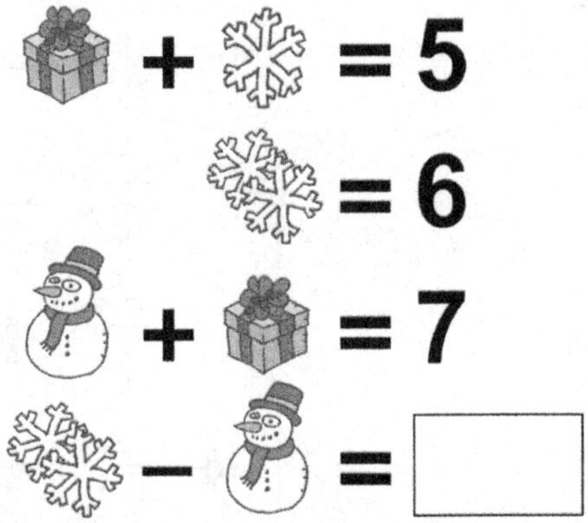

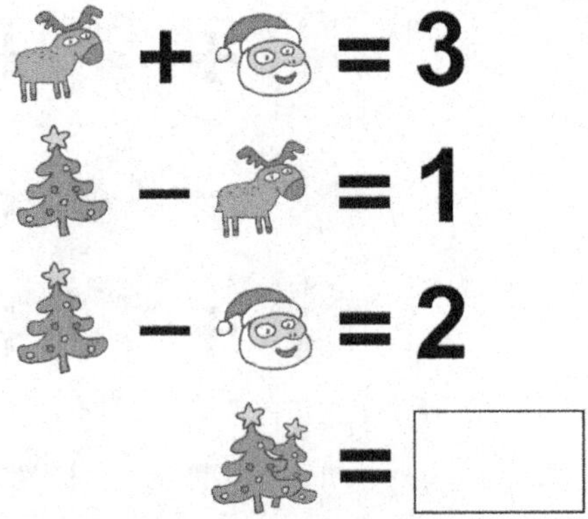

#31

🎅 + 🍬 = 13

🍬 + 🍬 + 🍬 = 9

🍬 − ❄️ + 🎅 = 7

❄️ = ☐

#32

🎁 − 🧝 = 2

🧝 − 🎄 = 3

🎄 + 🎄 = 9

🎁 + 🎁 + 🎁 = ☐

#33

 +  +  = 6

🦌 + HO = 4

☃️ − HO + 🦌 = 11

☃️ − 🦌 = ☐

15

#34

$$\text{☃} - \text{HO} = 2$$
$$\text{HO} + \text{HO} = 4$$
$$\text{🎄} + \text{HO} - \text{☃} = 1$$
$$\text{🎄} + \text{🎄} + \text{🎄} = \boxed{}$$

#35

$$\text{🦌} + \text{🦌} + \text{🦌} = 15$$
$$\text{🎅} - \text{🦌} = 2$$
$$\text{🎅} + \text{❄} = 15$$
$$\text{❄} + \text{❄} + \text{🦌} = \boxed{}$$

#36

$$\text{🎁} + \text{🍬} = 7$$
$$\text{🍬} + \text{🍬} + \text{🍬} = 9$$
$$\text{🧝} + \text{🎁} + \text{🧝} = 14$$
$$\text{🧝} - \text{🍬} = \boxed{}$$

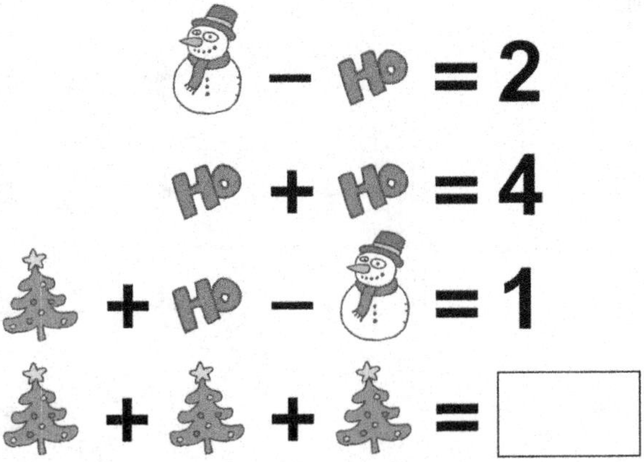

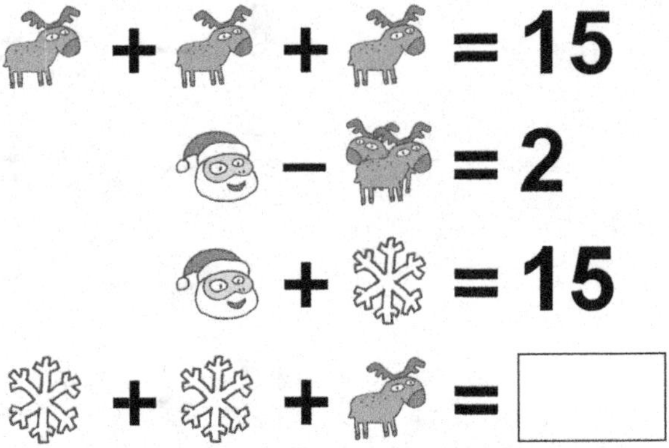

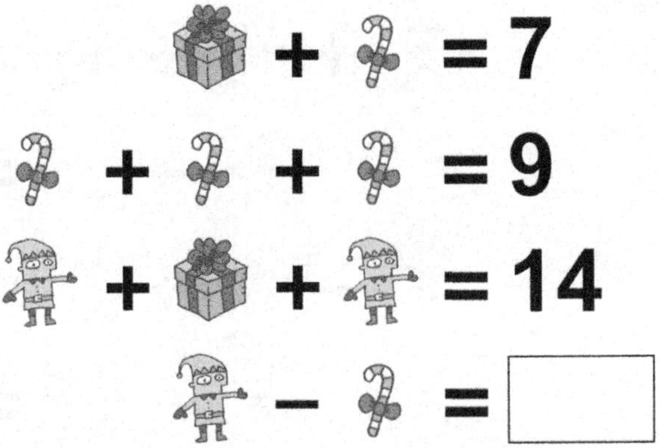

#37

$$\text{(snowmen)} + \text{(snowmen)} = 12$$
$$\text{(snowman)} + \text{(gift)} - \text{(snowflake)} = 1$$
$$\text{(gift)} + \text{(gift)} + \text{(gift)} = 15$$
$$\text{(snowflakes)} - \text{(snowman)} = \boxed{}$$

#38

$$\text{(candy cane)} + \text{(candy cane)} + \text{(candy cane)} = 18$$
$$\text{(candy canes)} - \text{(elf)} = 4$$
$$\text{HO} + \text{(elf)} = 11$$
$$\text{HOHOHO} - \text{(candy cane)} = \boxed{}$$

#39

$$\text{(Santa)} - \text{(reindeer)} + \text{(tree)} = 4$$
$$\text{(Santas)} - \text{(Santa)} = 5$$
$$\text{(tree)} + \text{(tree)} + \text{(tree)} = 9$$
$$\text{(Santa)} + \text{(reindeer)} - \text{(tree)} = \boxed{}$$

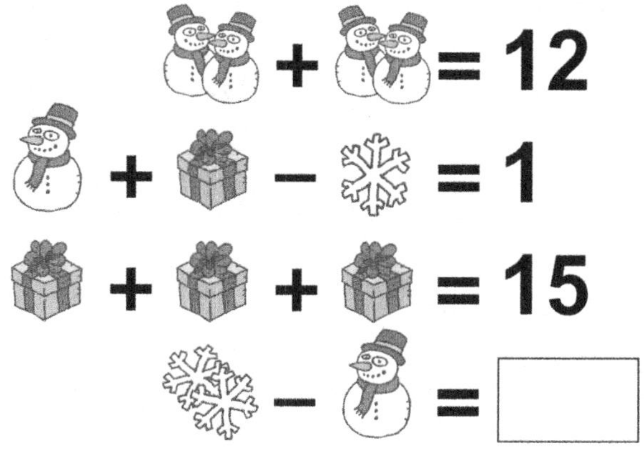

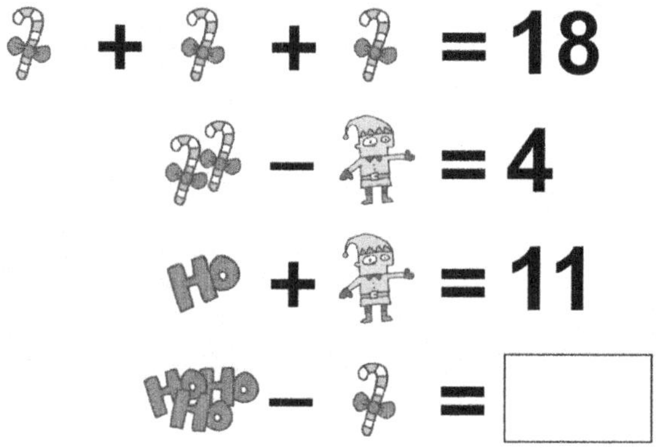

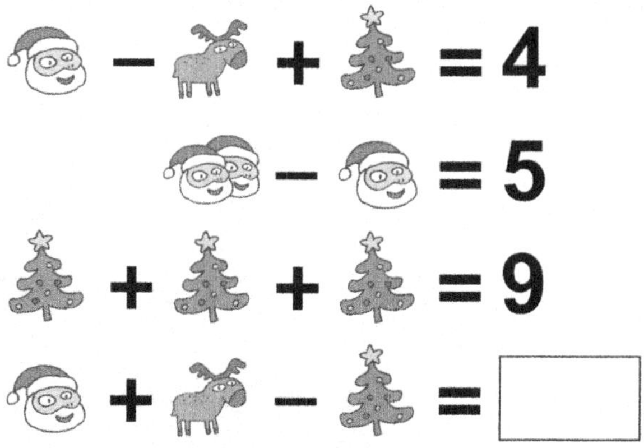

#40

НГо + ❄ = 8

❄ + ❄ = 8

🎄 − ❄ − ❄ = 1

НГо + НР + 🎄 = ☐

#41

🎅 + 🦌 = 11

🦌 = 4

🎅 + 🧝 = 8

🎅🎅 − 🧝 = ☐

#42

⛄ = 9

🎁 − ⛄ = 1

⛄ + 🎁 + 🍬 = 18

🎁 − 🍬 = ☐

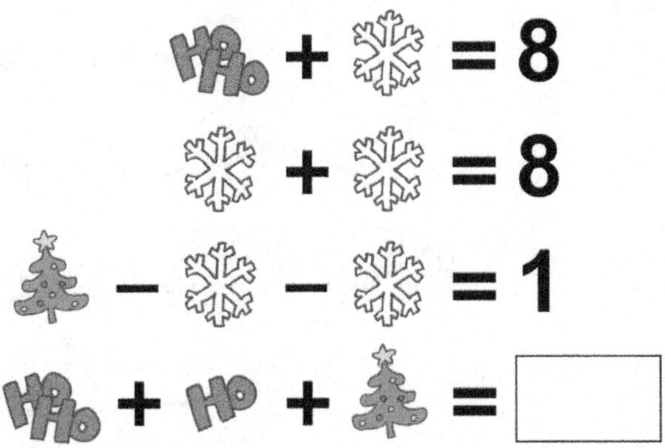

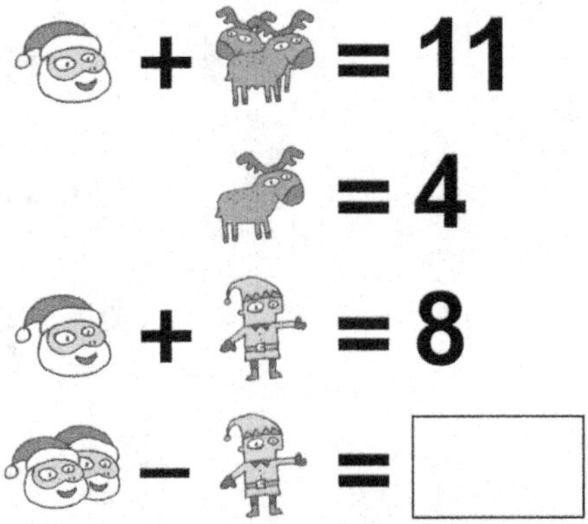

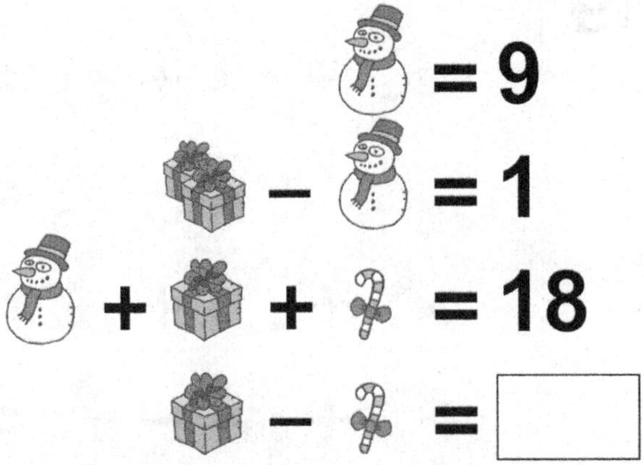

#43

$$\text{🎅} - \text{🍬} = 2$$

$$\text{🍬} = 5$$

$$\text{🧝} - \text{⛄} + \text{🍬} = 8$$

$$\text{🍬} - \text{🧝} + \text{⛄} = \boxed{}$$

#44

$$\text{🎄} + \text{🎁} - \text{HO} = 6$$

$$\text{HO} + \text{HO} + \text{HO} = 21$$

$$\text{🎁} + \text{🎁} + \text{HO} = 27$$

$$\text{HOHO} - \text{🎄} - \text{🎁} = \boxed{}$$

#45

$$\text{🎅} + \text{🦌} = 20$$

$$\text{🦌} + \text{🦌} + \text{🦌} = 3$$

$$\text{🎅} - \text{❄} + \text{🦌} = 5$$

$$\text{❄} - \text{🦌} + \text{🎅} = \boxed{}$$

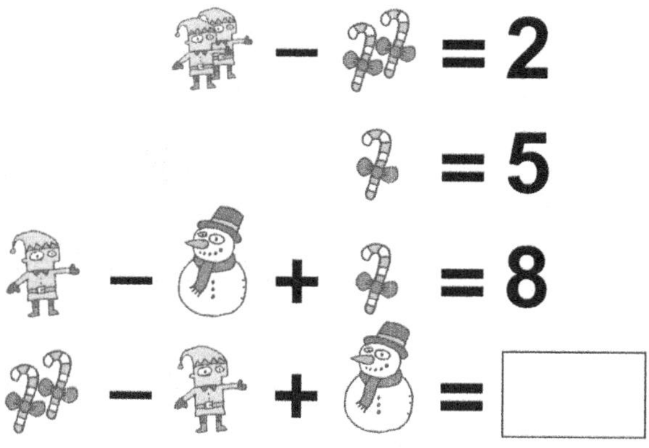

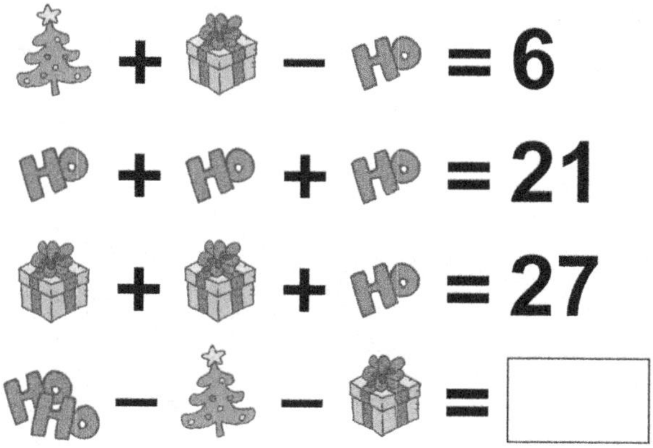

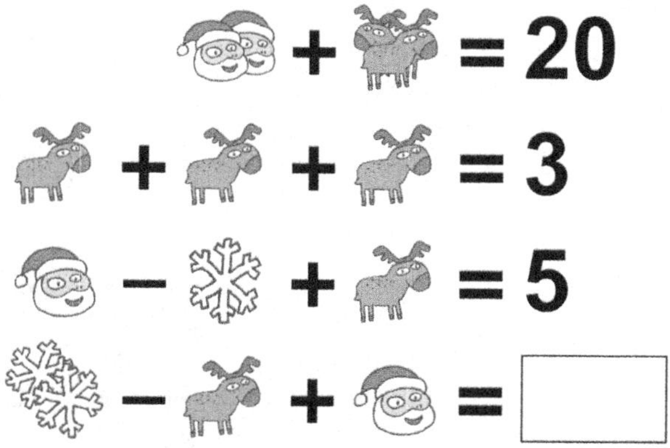

#46

🎅🎅 + ⛄ = 3

HO + ⛄ = 5

🎄 − HO − HO = 1

🎄 + ⛄ =

#47

❄ − 🍬 = 2

🍬🍬 = 6

❄ + ❄ − 🧝 = 7

🍬 + 🧝 =

#48

🎁 = 7

🎁🎁 + 🎅 = 20

🦌 − 🎅 = 5

🦌 + 🎅 − 🎁 =

20

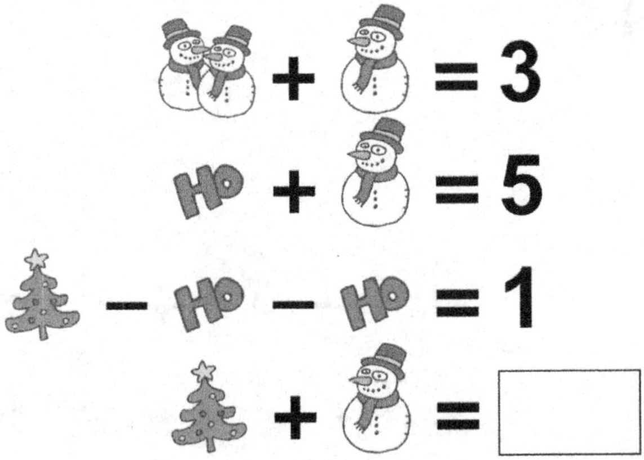

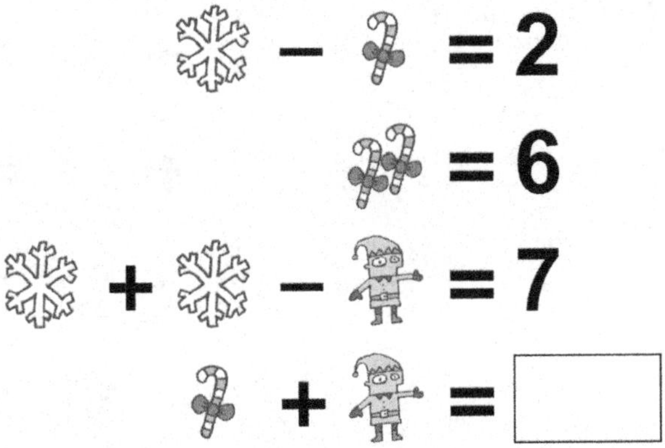

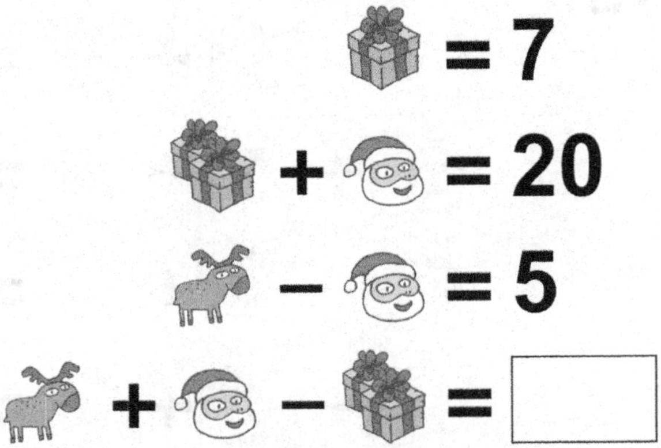

#49

#50

#51

21

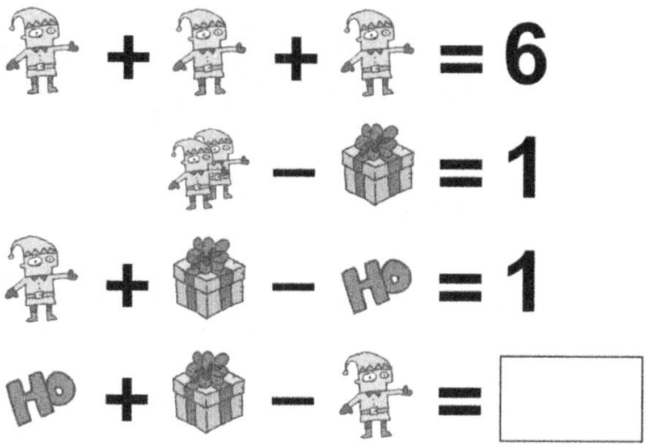

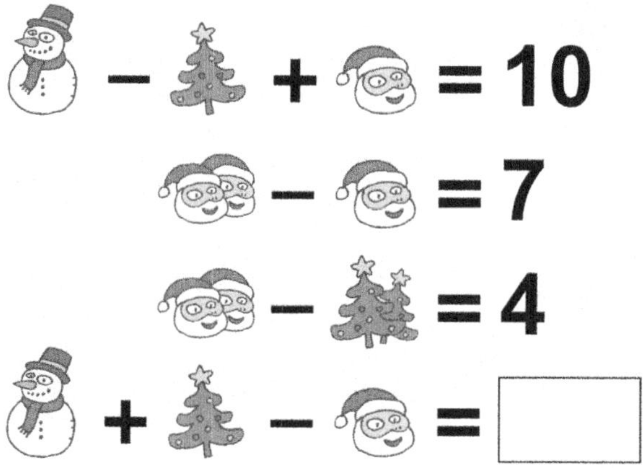

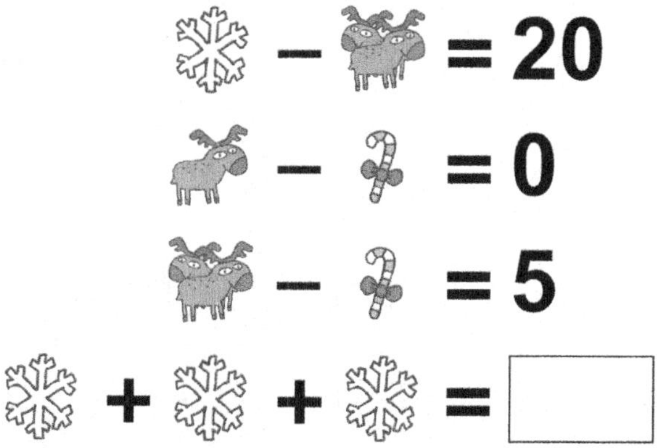

#52

🎁 − 🍬 = 2

🎁 − 🍬 = 5

❄ − 🎁 = 3

❄ + 🍬 = ☐

#53

🦌 + 🦌 + 🦌 = 16

🎅 + 🎅 − 🦌 = 6

🎄 − 🎅 = 2

🎄 − 🦌 = ☐

#54

Ho + Ho = 4

🧝 − HoHo = 1

HoHo + 🧝 + ⛄⛄ = 20

Ho + 🧝 − ⛄ = ☐

22

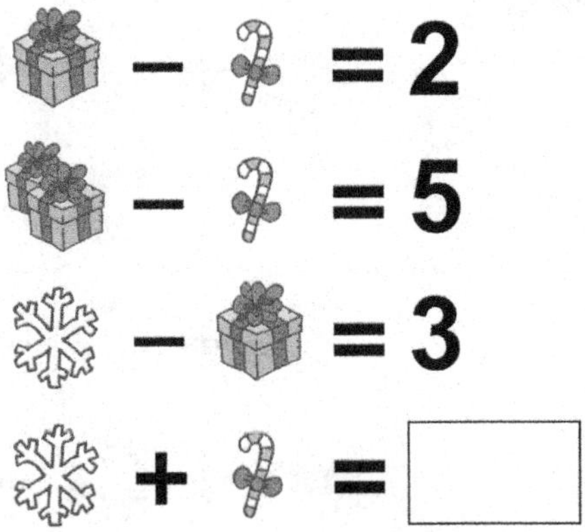

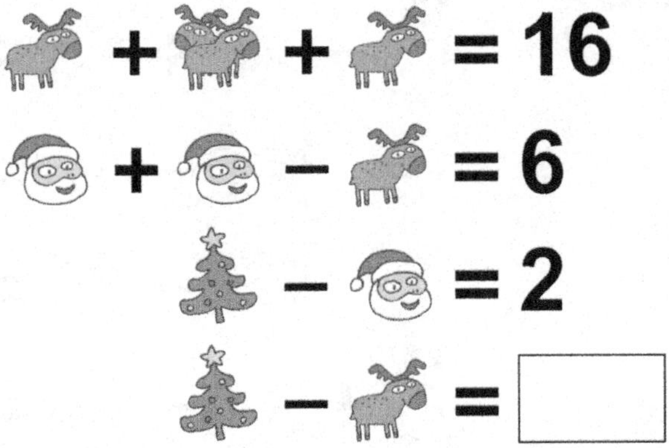

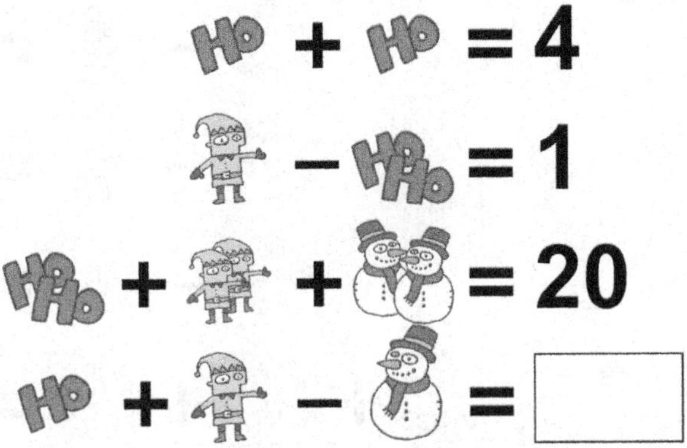

#55

🎄 + 🦌 − HP = 20

HP + HP + HP = 60

🦌 − 🦌 + HP = 30

🎄 + 🎄 = ☐

#56

🎁 + ❄️ − 🍬 = 11

❄️ = 9

❄️ + 🍬 = 21

🍬 + 🎁 + 🎁 = ☐

#57

⛄ − 🎅 = 1

🎅 + 🎅 + 🎅 = 3

⛄ + 🧝 = 9

🧝 − ⛄ − 🎅 = ☐

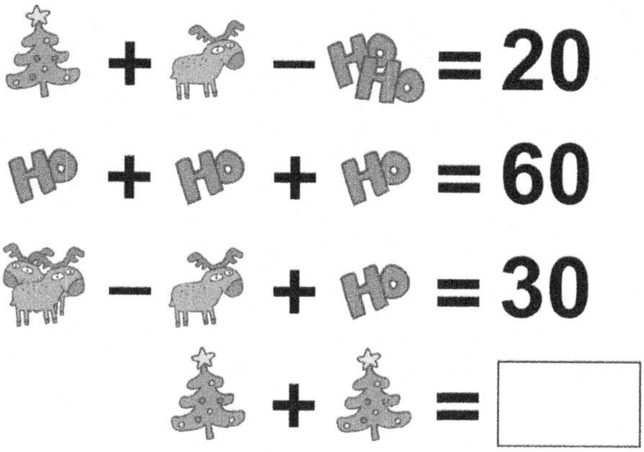

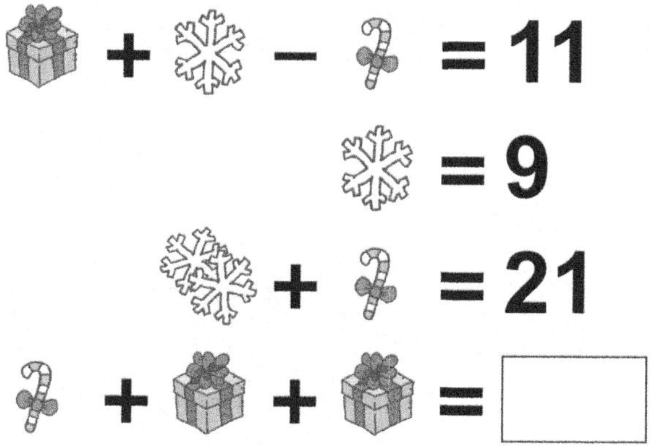

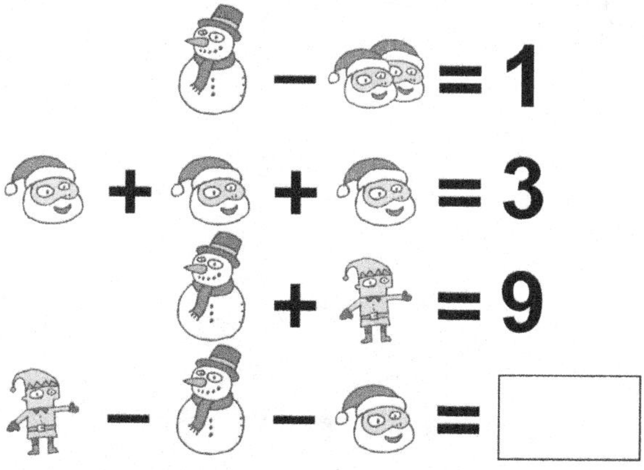

#58

🍬 − 🎁 − 🧝 = 1

🧝 = 2

🎁 + 🎁 = 6

🍬 + 🎁 + 🧝 = ⬜

#59

🦌 + 🦌 + ❄️ = 26

❄️ = 3

🦌 − HO = 1

HO + HO + HO = ⬜

#60

🎅 + 🎄 = 15

🎄 − 🎄 = 7

🎄 + ⛄ + 🎅 = 18

⛄ + ⛄ − 🎅 = ⬜

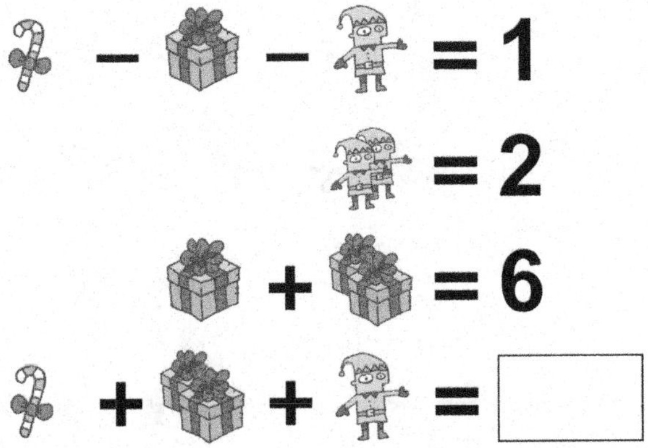

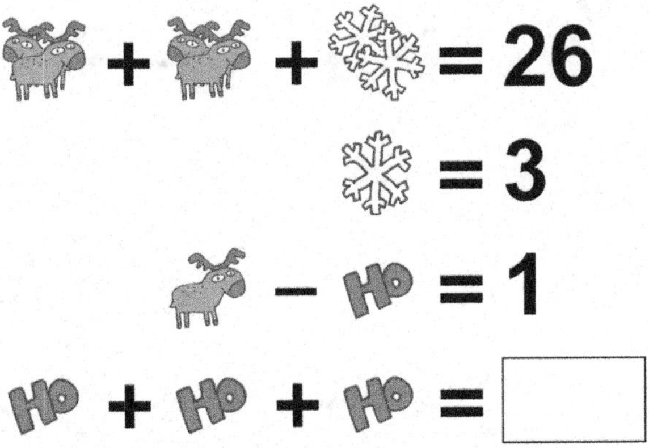

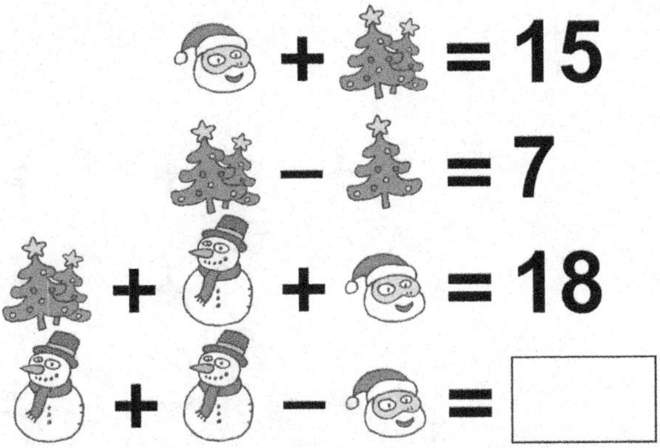

#61

#62

#63

25

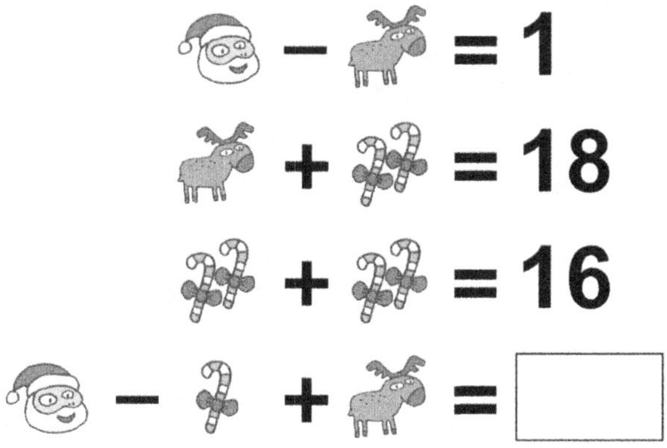

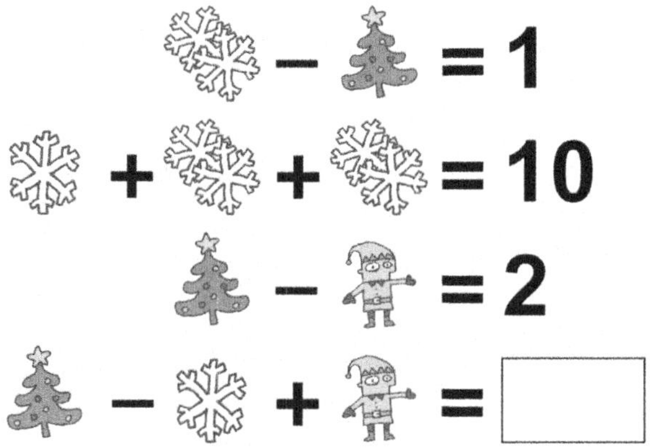

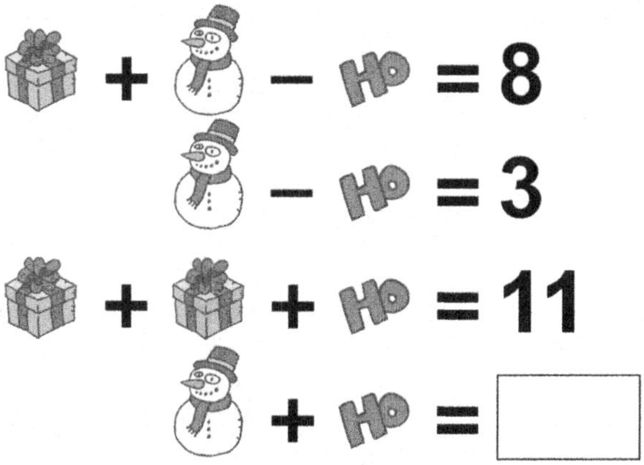

#64

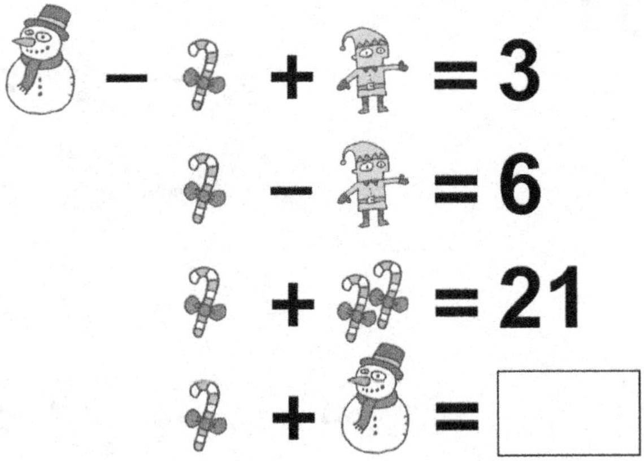

snowman − candycane + elf = 3

candycane − elf = 6

candycane + candycane = 21

candycane + snowman = ☐

#65

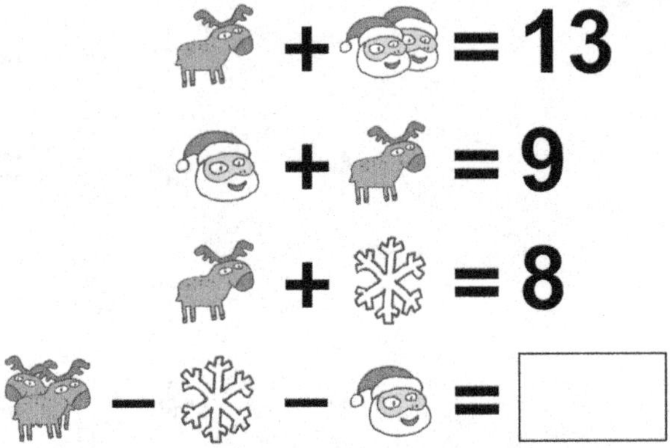

reindeer + santa = 13

santa + reindeer = 9

reindeer + snowflake = 8

reindeer − snowflake − santa = ☐

#66

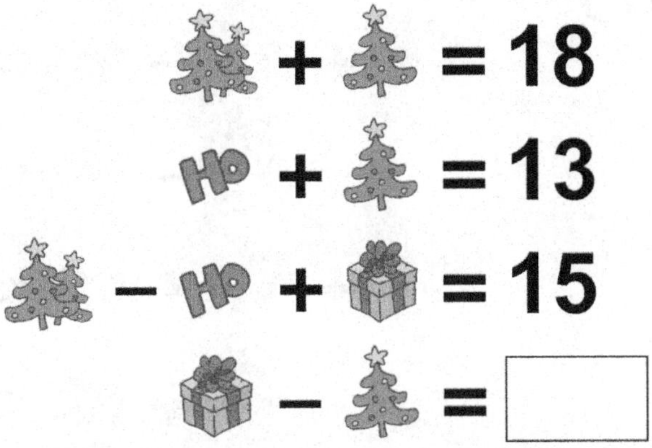

tree + tree = 18

HO + tree = 13

tree − HO + gift = 15

gift − tree = ☐

26

#67

$$HO + HO = 8$$
$$HO + \text{(moose)} + \text{(santa)} = 20$$
$$HO + \text{(moose)} + \text{(moose)} = 14$$
$$\text{(santa)} + \text{(santa)} + \text{(santa)} = \boxed{}$$

#68

$$\text{(elf)} + \text{(snowman)} = 5$$
$$\text{(candy)} + \text{(elf)} + \text{(snowman)} = 9$$
$$\text{(candy)} + \text{(elf)} = 7$$
$$\text{(snowman)} + \text{(snowman)} = \boxed{}$$

#69

$$\text{(tree)} + \text{(tree)} + \text{(gift)} = 10$$
$$\text{(tree)} + \text{(tree)} + \text{(gift)} = 16$$
$$\text{(gift)} - \text{(tree)} + \text{(snowflake)} = 9$$
$$\text{(snowflake)} + \text{(tree)} - \text{(gift)} = \boxed{}$$

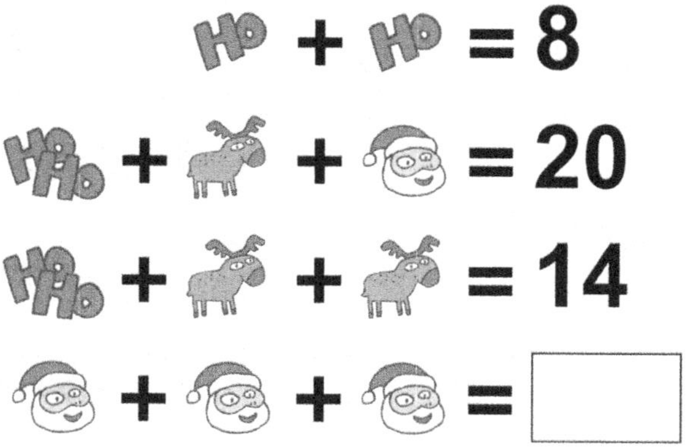

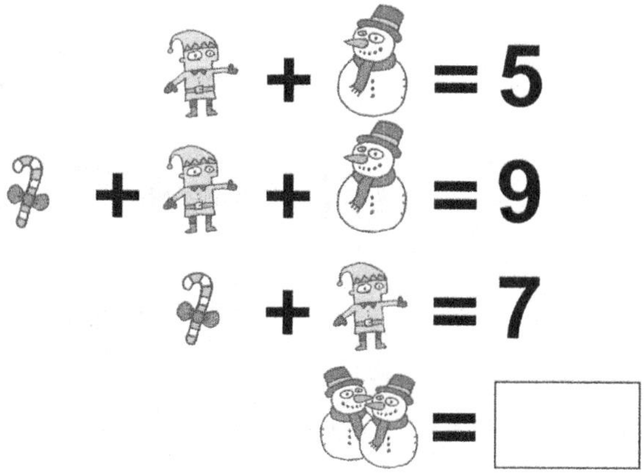

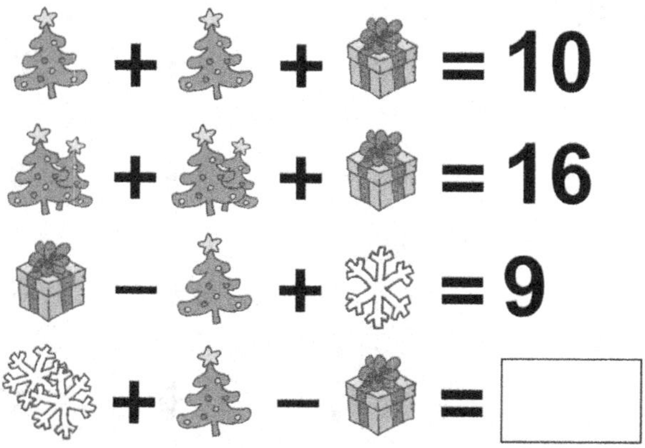

#70

$$\text{🍬} + \text{🍬🍬} - \text{🎁} = 7$$

$$\text{HO} - \text{🎁} + \text{🎁} = 10$$

$$\text{🍬🍬} = 8$$

$$\text{🎁} + \text{🍬} + \text{HO} = \boxed{}$$

#71

$$\text{🎅} - \text{🎄} = 1$$

$$\text{🎅} + \text{🎅} + \text{🎅} = 15$$

$$\text{🎅} - \text{❄} = 1$$

$$\text{❄} - \text{🎄} + \text{🎅} = \boxed{}$$

#72

$$\text{🦌} - \text{⛄} + \text{⛄} = 7$$

$$\text{🦌} + \text{⛄} + \text{⛄} = 15$$

$$\text{⛄⛄} - \text{🧝} + \text{🦌} = 11$$

$$\text{🧝} + \text{⛄} + \text{🦌} = \boxed{}$$

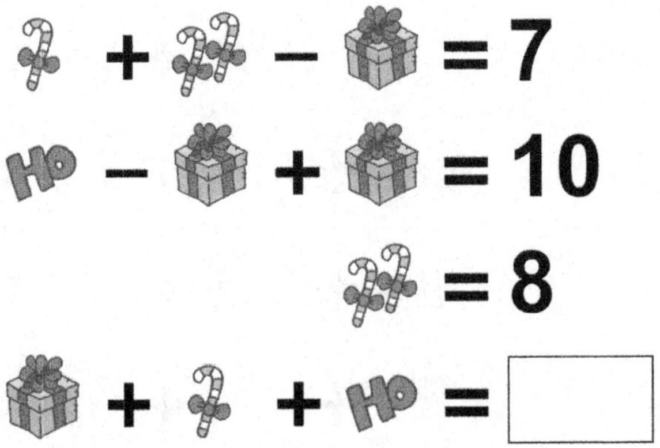

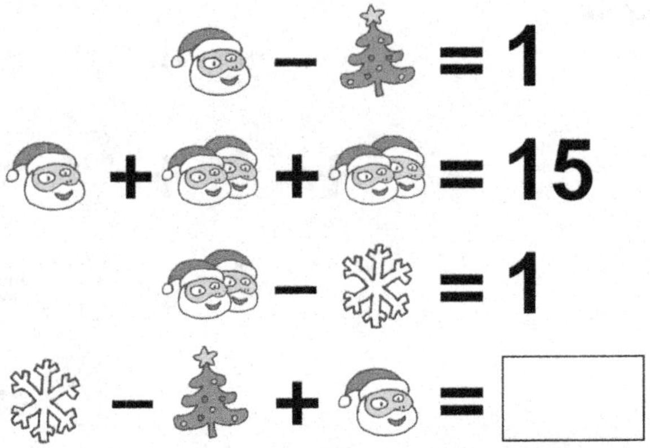

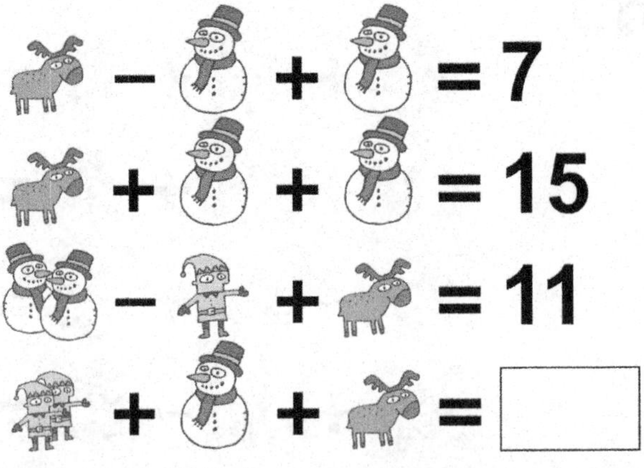

#73

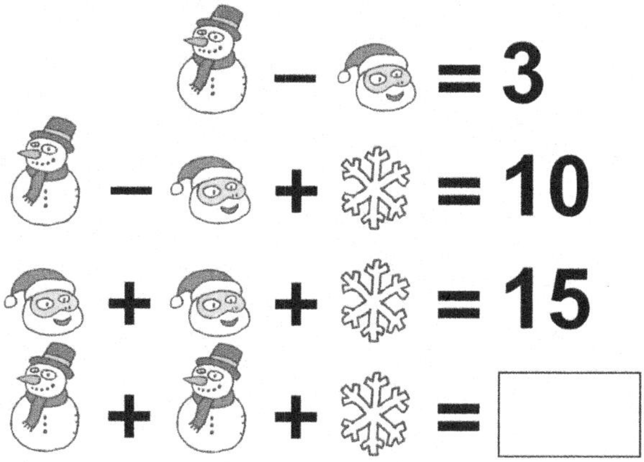

☃ − 🎅 = 3

☃ − 🎅 + ❄ = 10

🎅 + 🎅 + ❄ = 15

☃ + ☃ + ❄ = ☐

#74

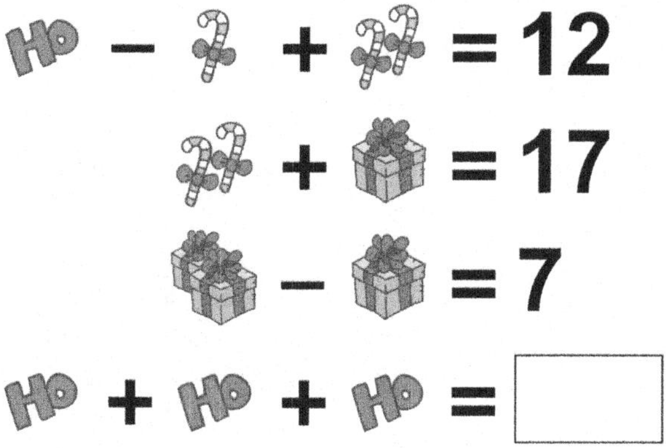

HO − 🍬 + 🍬 = 12

🍬 + 🎁 = 17

🎁 − 🎁 = 7

HO + HO + HO = ☐

#75

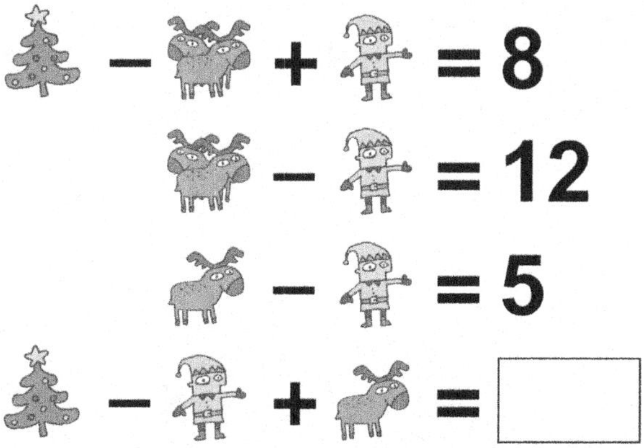

🎄 − 🦌 + 🧝 = 8

🦌 − 🧝 = 12

🦌 − 🧝 = 5

🎄 − 🧝 + 🦌 = ☐

#76

 − = **3**

 = **9**

 − = **1**

 + + = ☐

#77

 +  = **9**

 −  = **7**

 +  = **3**

 +  = ☐

#78

 + − = **13**

 −  = **7**

 − = **3**

 +  + = ☐

#79

🎅 + 🦌 = 101

❄️ + 🎅 = 99

❄️ + 🦌 = 4

🎅 − ❄️ − 🦌 = ☐

#80

H₂O − 🍬 + H₂O = 18

🍬 + 🍬 = 24

🎅 − H₂O = 35

🎅 − 🍬 = ☐

#81

🧝 + 🎁 + 🔥 = 15

🧝 + 🎁 = 6

🔥 − 🧝 = 5

🎁 + 🎁 = ☐

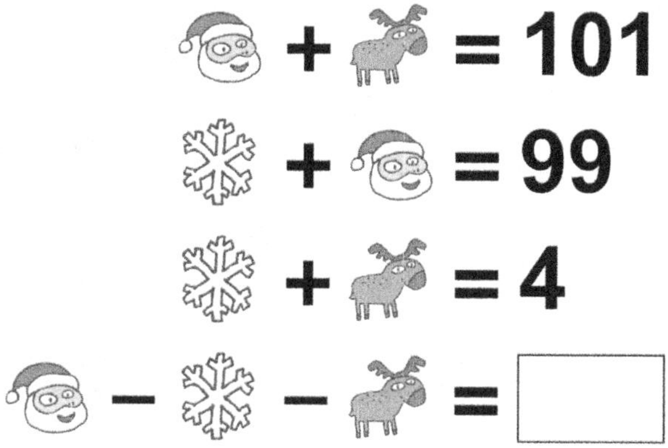

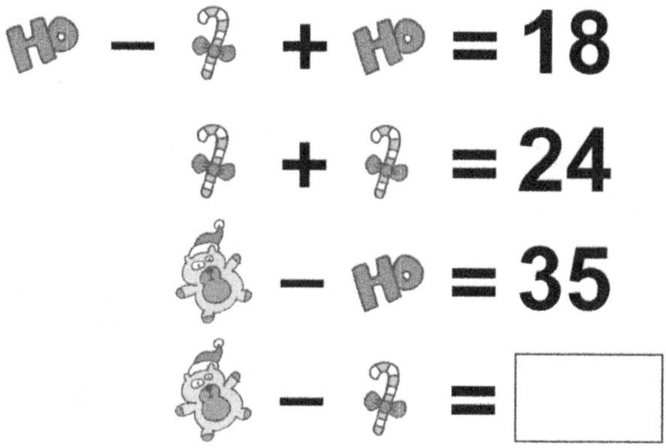

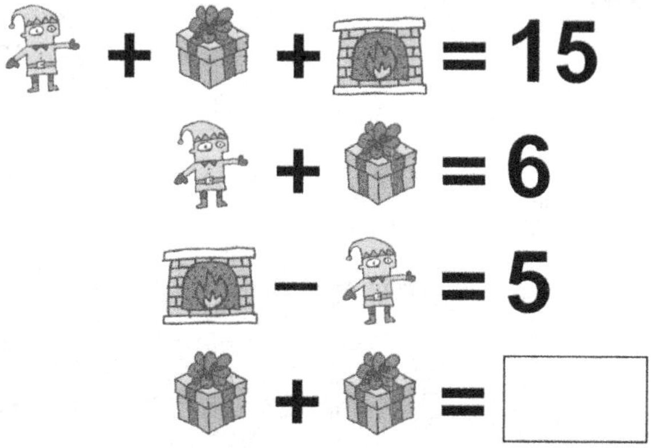

#82

#83

#84

🎀 + 🎀 = 6

🍬 − 🎀 = 4

🍬 + 🍬 + 🧝 = 40

🧝 + 🧝 + 🎀 = ▢

32

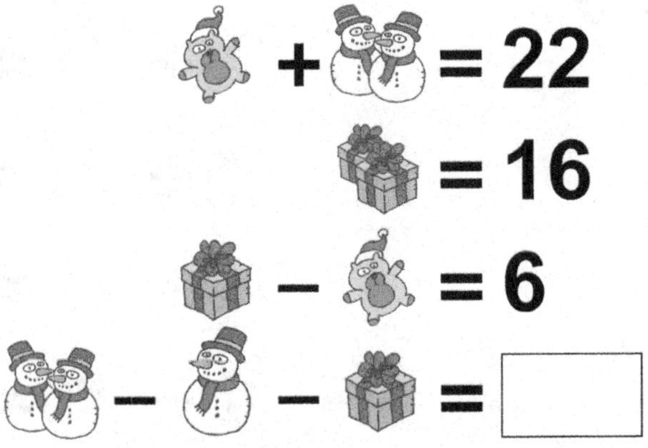

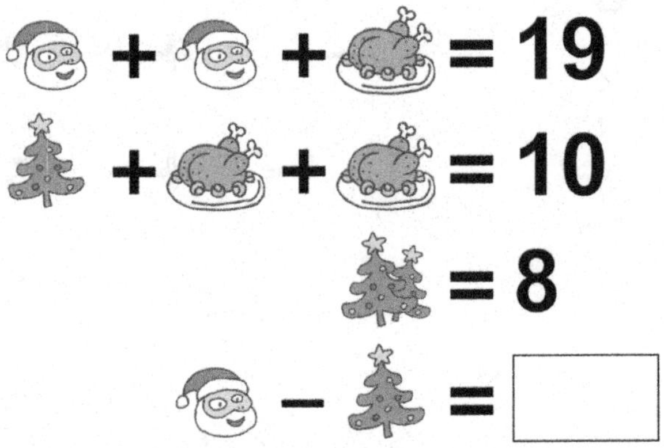

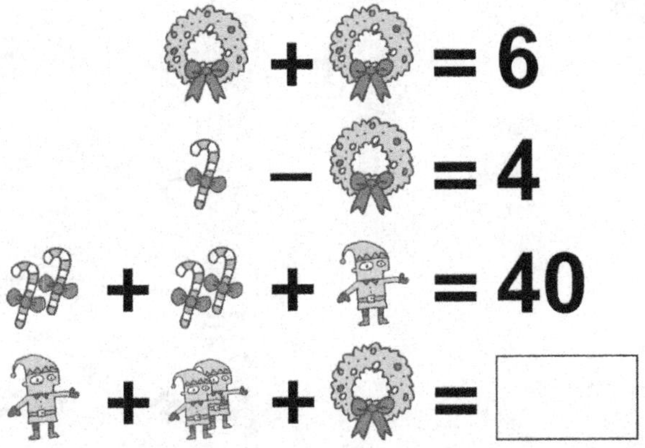

#85

snowman − snowglobe = 1

snowman − wreath = 25

snowglobe + snowglobe = 98

snowman + wreath + two snowmen = ☐

#86

elf + gift + gift = 24

gift − turkey = 3

elf + gift = 17

elf + turkey + turkey = ☐

#87

two characters − tree = 13

character − tree = 4

character + HO = 15

tree + HOHO = ☐

33

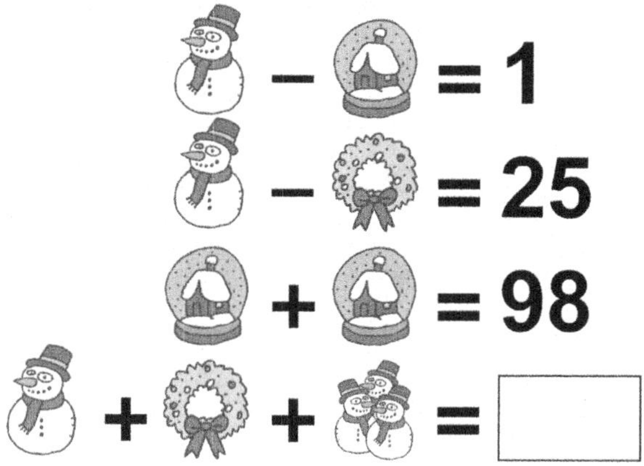

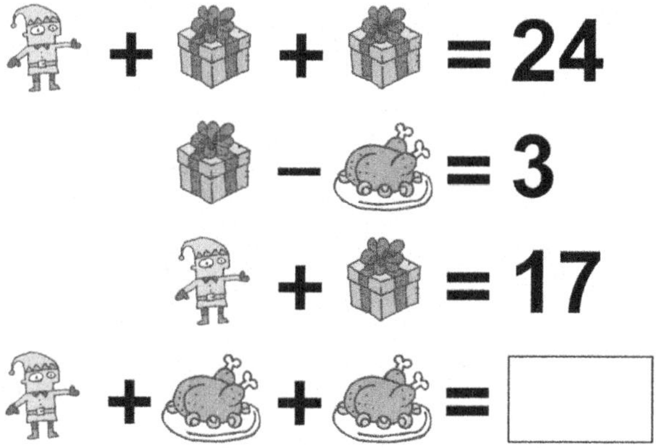

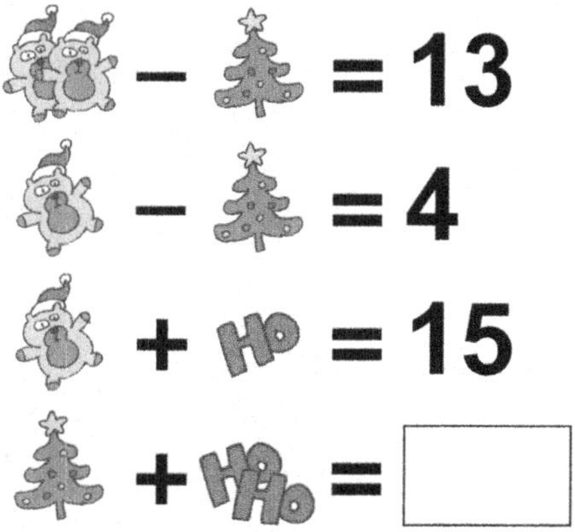

#88

HO + HOHO + ❄ = 38

HO − ❄ + 🎅 = 11

❄ = 7

HO + 🎅 + 🎅 = ☐

#89

🦌 + 🎁 − ⛄ = 7

🦌 − ⛄ = 4

🎁 + 🦌 = 9

⛄ + 🦌 + 🦌 = ☐

#90

🎅 − 🎅 = 6

🎅 − 🎄 + 🔥 = 7

🔥 = 5

🎄 + 🎅 + 🎅 = ☐

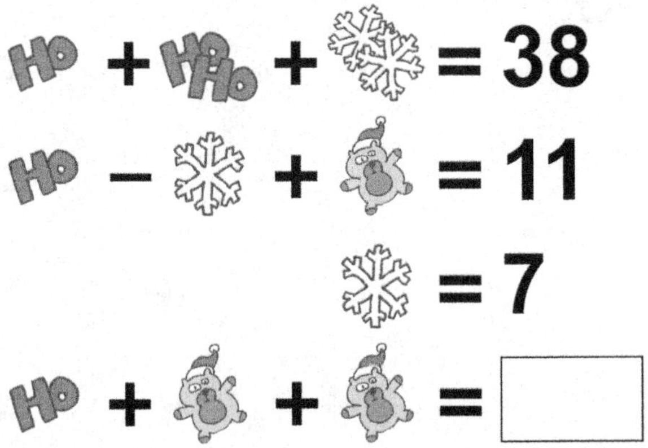

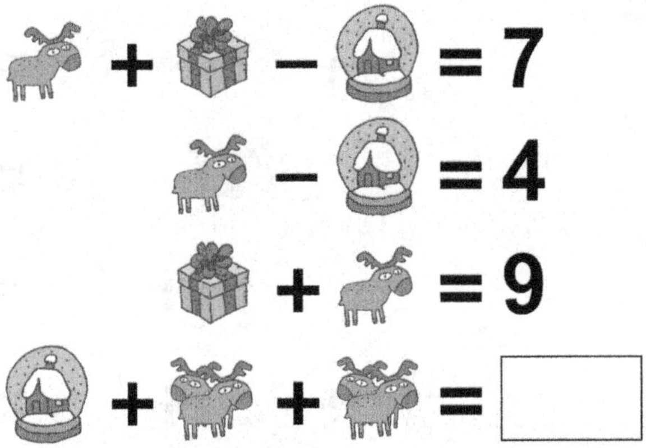

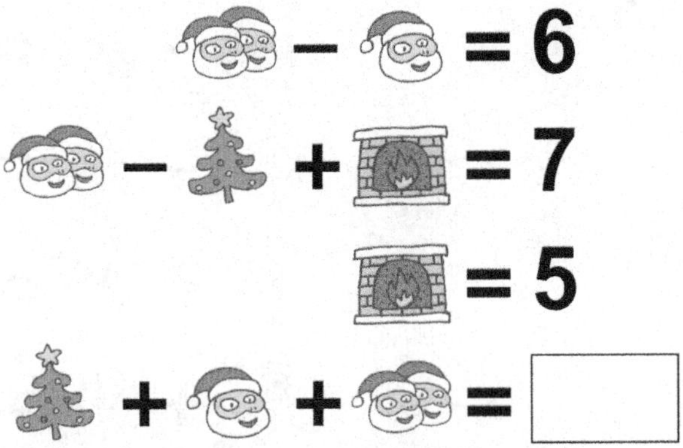

#91

🍬 + 🎀 = 20

🎁 + 🍬 = 57

🎁 + 🎁 = 100

🎀 − 🍬 + 🎁 = ☐

#92

#93

35

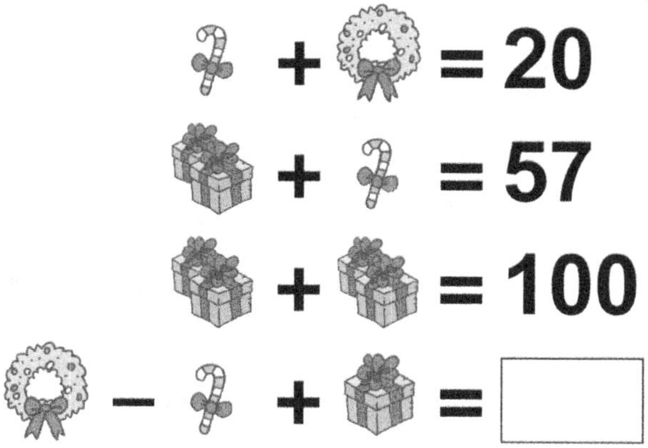

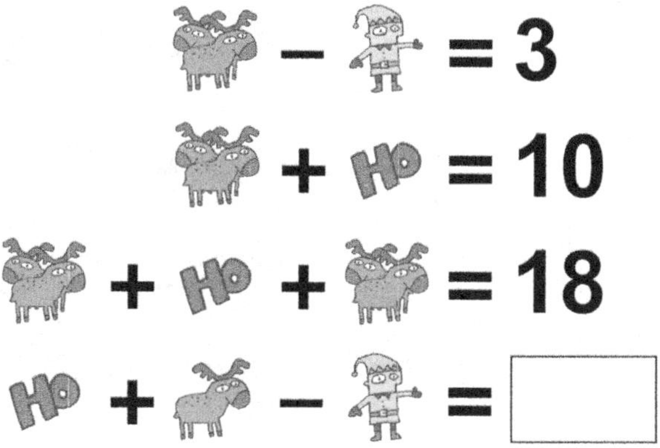

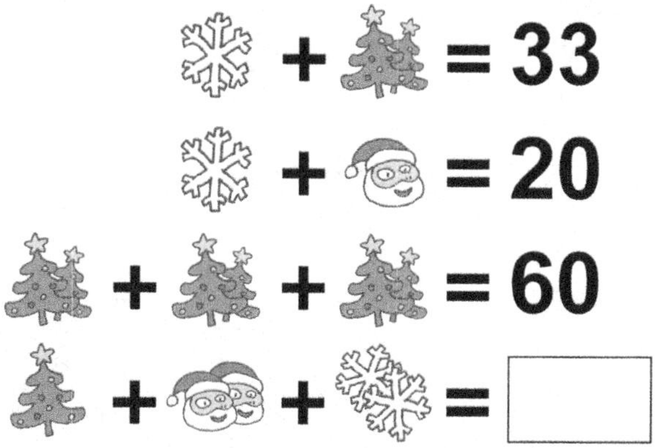

#94

$$\text{🦌} - \text{🍬} = 3$$
$$\text{🦌} + \text{🦌} + \text{🦌} = 30$$
$$\text{🍬🍬} - \text{🎅} = 1$$
$$\text{🎅} + \text{🍬} = \boxed{}$$

#95

$$\text{❄} - \text{🎁} = 2$$
$$\text{❄} + \text{🍗} + \text{🍗} = 18$$
$$\text{❄❄} - \text{🎁} = 6$$
$$\text{🍗} - \text{🎁} = \boxed{}$$

#96

$$\text{⛄} + \text{⛄⛄} + \text{🎄} = 18$$
$$\text{🎄} + \text{HO} - \text{🎄} = 10$$
$$\text{HOHO} = 14$$
$$\text{⛄⛄} - \text{🎄} = \boxed{}$$

36

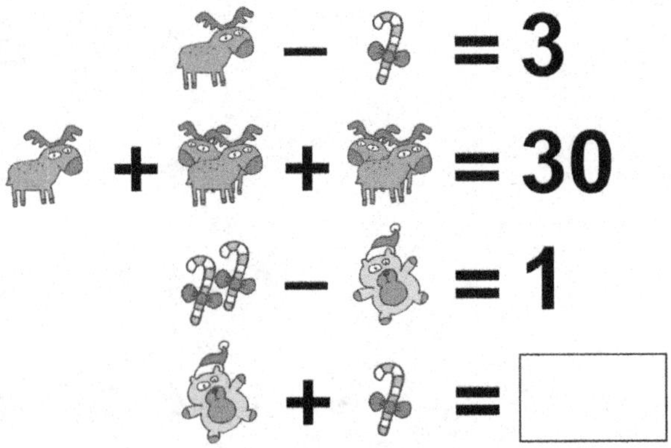

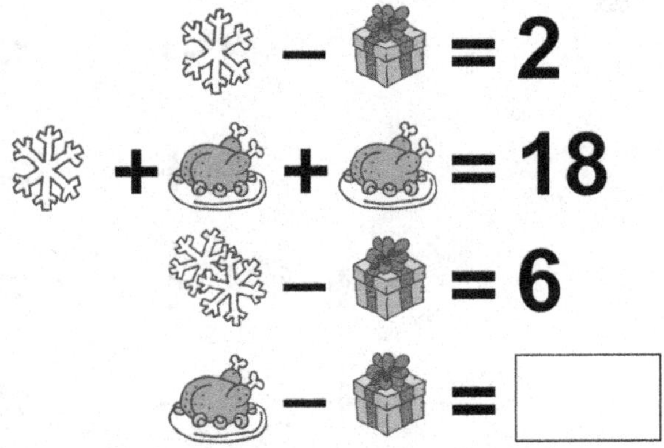

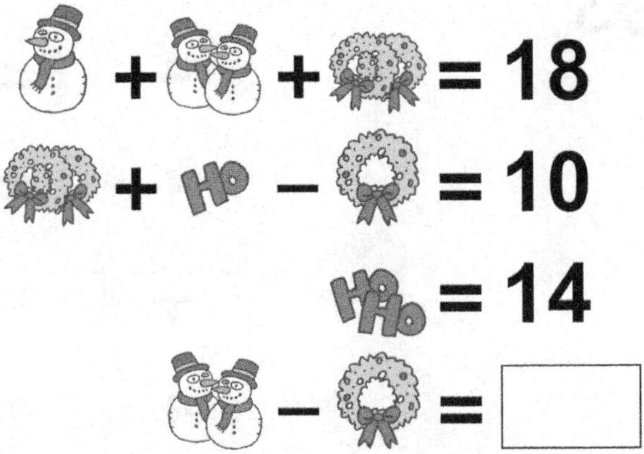

#97

$$\text{🍗} - \text{🔥} = 3$$
$$\text{❄} + \text{🍗} - \text{🔥} = 8$$
$$\text{🍗} + \text{🍗} - \text{🔥} = 7$$
$$\text{❄} + \text{🍗} + \text{🔥} = \boxed{}$$

#98

$$\text{🎄} + \text{🎅} = 26$$
$$\text{🎅} + \text{🎄} = 21$$
$$\text{🦌} - \text{🎄} = 1$$
$$\text{🎅} + \text{🎄} + \text{🦌} = \boxed{}$$

#99

$$\text{🎅} + \text{🎅} = 100$$
$$\text{⛄} - \text{🎅} = 10$$
$$\text{⛄} - \text{🎅} + \text{🧝} = 9$$
$$\text{⛄} - \text{🧝} = \boxed{}$$

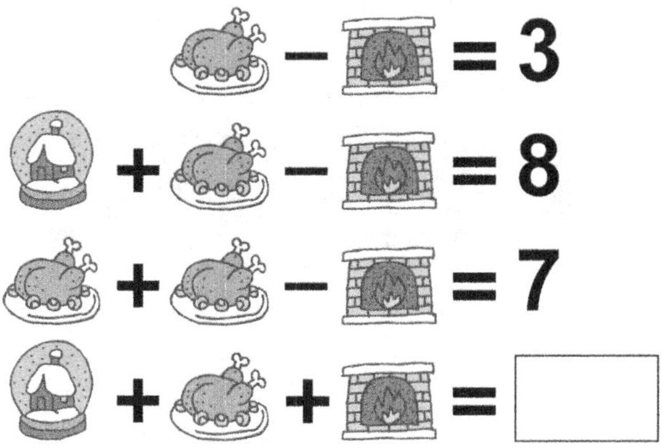

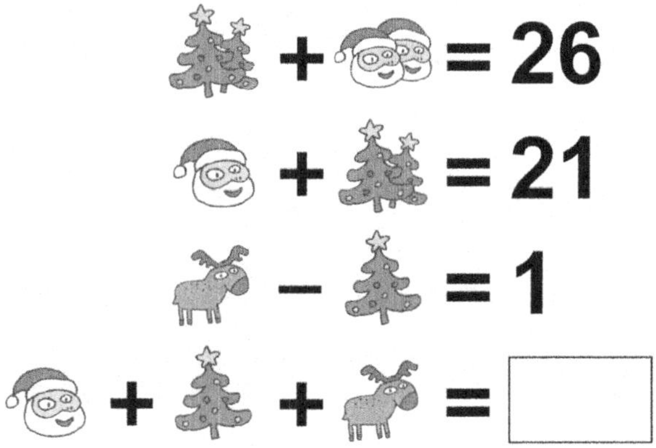

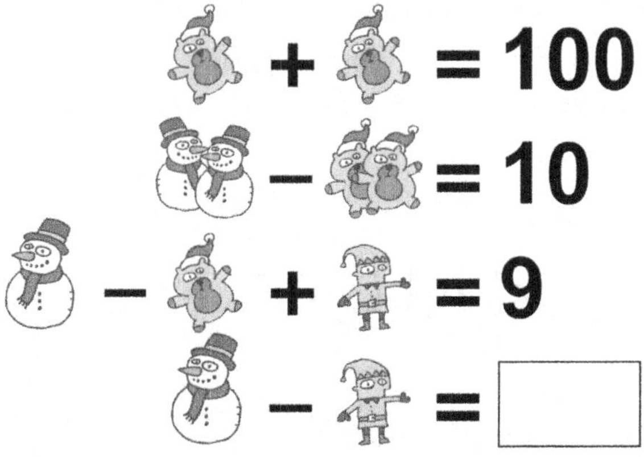

LEVEL 2

The puzzles in Level 2 are slightly harder. You may have to study the equations longer and in different orders to figure out the value of each symbol. And as always, remember these rules...

1. Every symbol has a value of 1 or higher.

2. Two different symbols can have the same value.

3. There are no fractions or negative numbers.

4. Subtracting a larger number from a smaller number is not allowed.

5. The final answer is always 0 or higher.

Alright, let's keep going!

#100

 +   = 6

 −  = 2

 + + = ☐

#101

 −  = 5

 −  = 3

 +  =

 +  + = ☐

#102

 + +  = 9

 + + =

 −  = 5

 −  = ☐

#103

elf + elf = snowflake

snowflake − elf = 6

elf + gift = 9

gift + snowflake = ☐

#104

tree + candy = fireplace

tree + candy = 9

fireplace + fireplace + tree = 21

fireplace − candy = ☐

#105

turkey + snowmen + wreath = 30

turkey + wreath = snowman

turkey − wreath = 6

snowman − wreath = ☐

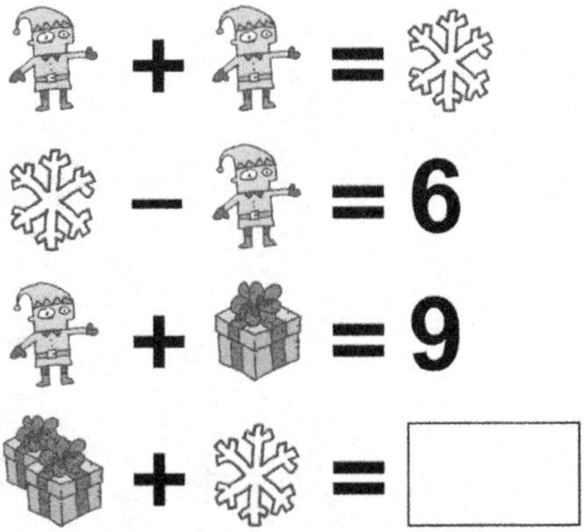

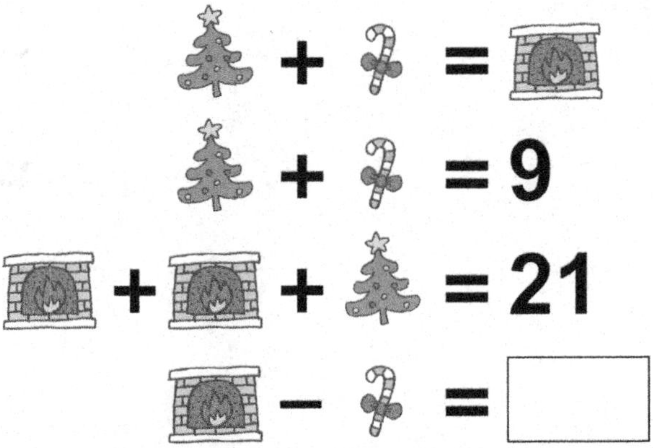

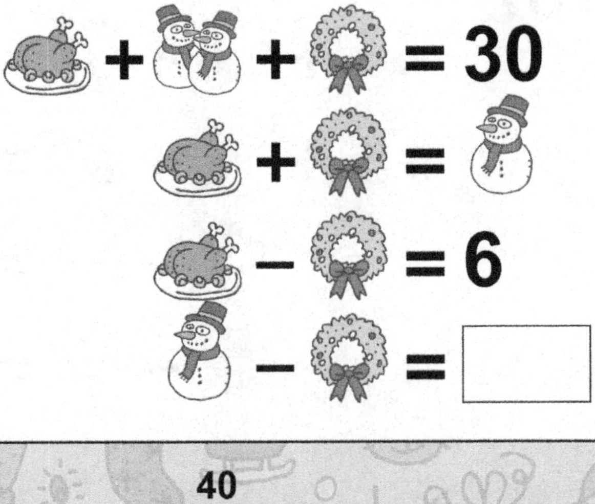

#106

🐷 = ☃☃

🐷 + 🐷 = **16**

🔥 + ☃ = **20**

🔥 − 🐷 = ☐

#107

Ho + 🍬 = **5**

Ho + 🔮 = **6**

Ho + HoHo = **12**

HoHo − 🍬 − 🔮 = ☐

#108

🐐 = **38**

🦌 + 🎄 = 🍗

🎄 + 🎄 = **22**

🍗 + 🍗 + 🍗 = ☐

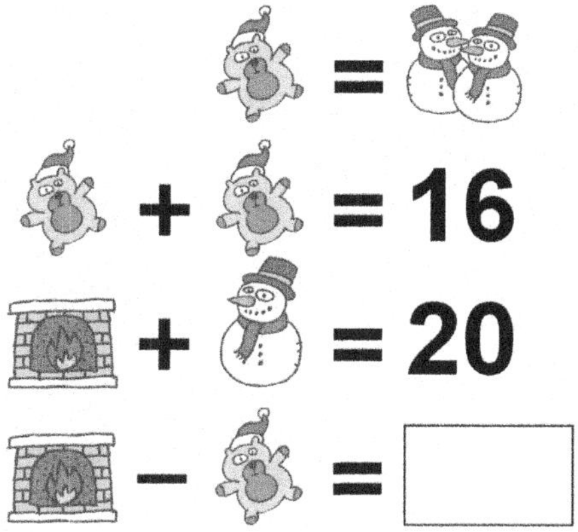

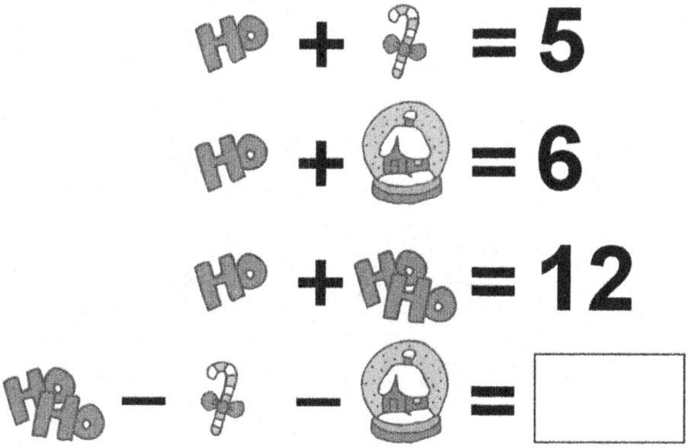

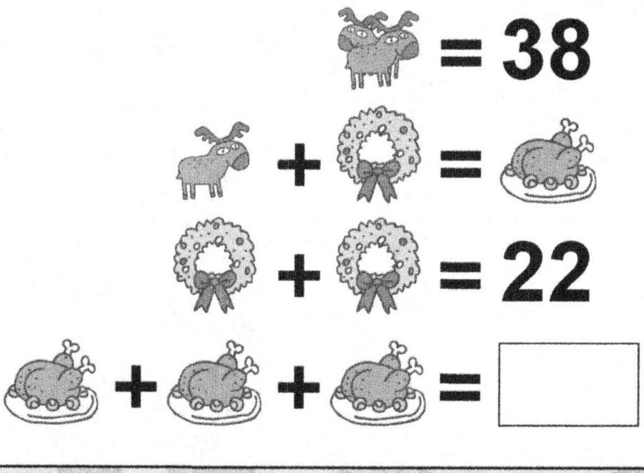

#109

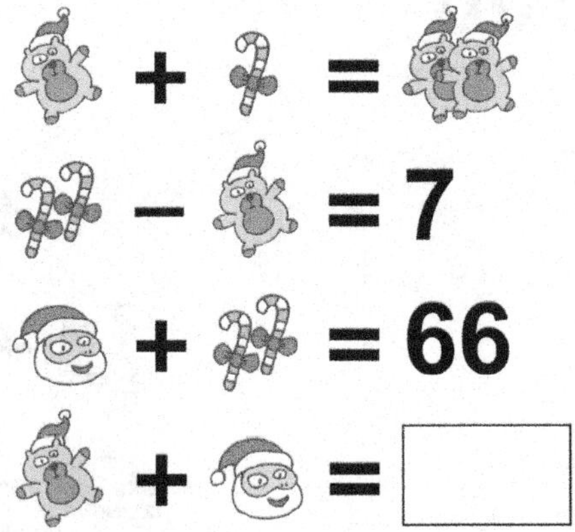

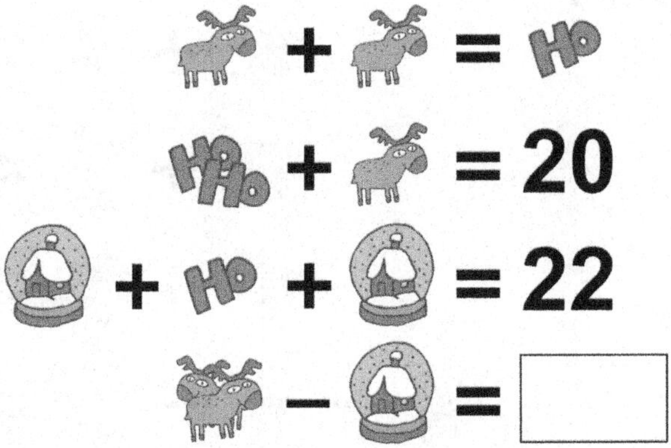

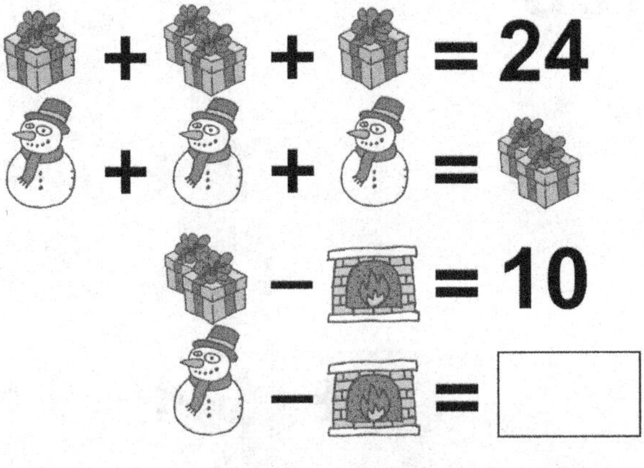

#110

#111

#112

❄ − 🏠 − 🏠 = 3

❄ − 🏠 = 4

❄ − 🍗 = 1

🍗 − 🏠 = ☐

#113

🎅 + 🎅 = 24

🐷 + 🎄 = 🎅🎅

🐷 + 🐷 + 🐷 = 9

🎄 − 🎅 = ☐

#114

🔮 − 🎄 = 1

🎄 + 🎄 = 🔮

🎁 − 🔮 = 3

🎁 + 🎄 = ☐

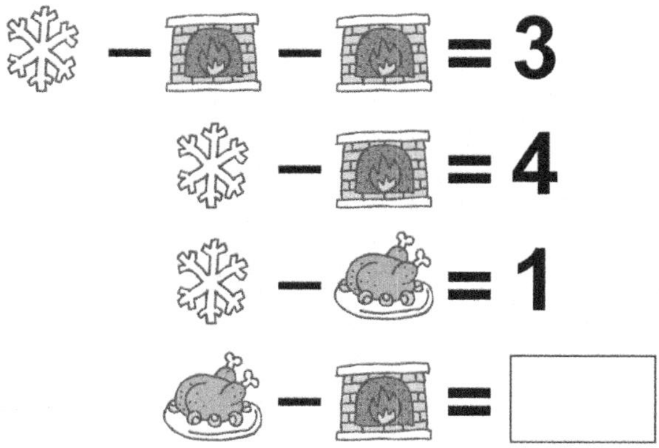

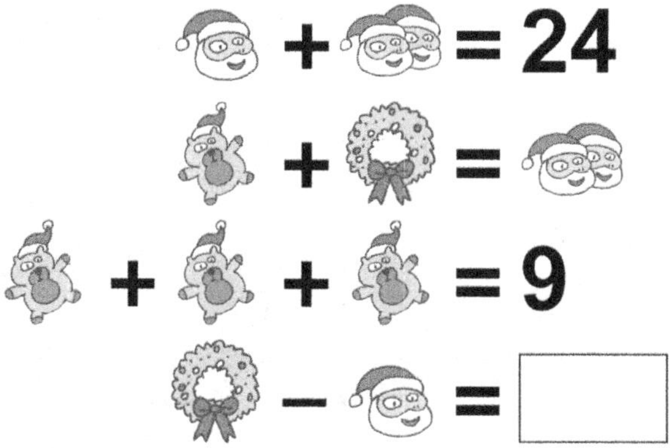

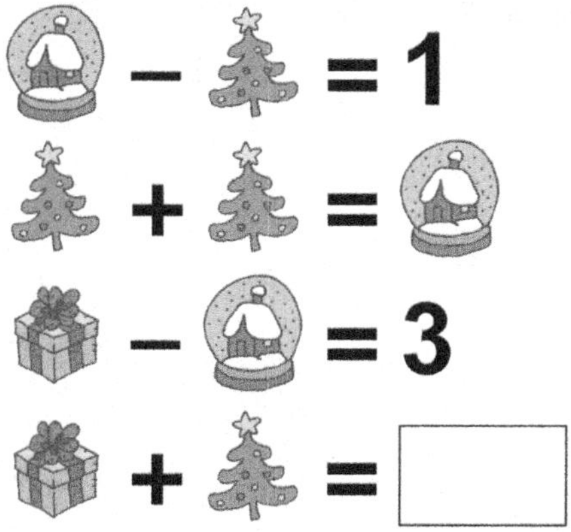

#115

🏠 − 👶 = 2

🏠 + 🏠 − 👶 = 8

👶 + 🧝 + 🧝 = 22

🧝 − 🏠 = ☐

#116

🎅 − 🔥 = 🔥

🎅 + 🔥 = 15

❄ + ❄ + ❄ = 42

❄ + ❄ + 🎅 = ☐

#117

🦌 + ⛄ + ⛄ = 7

🍬 − 🦌 = ⛄

🍬 − ⛄ = 5

🍬 + 🦌 = ☐

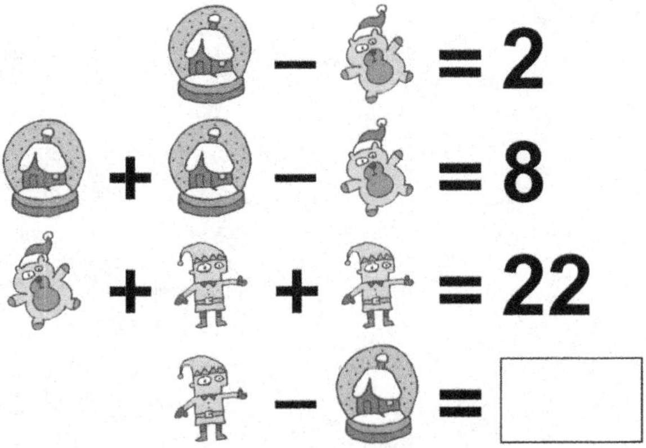

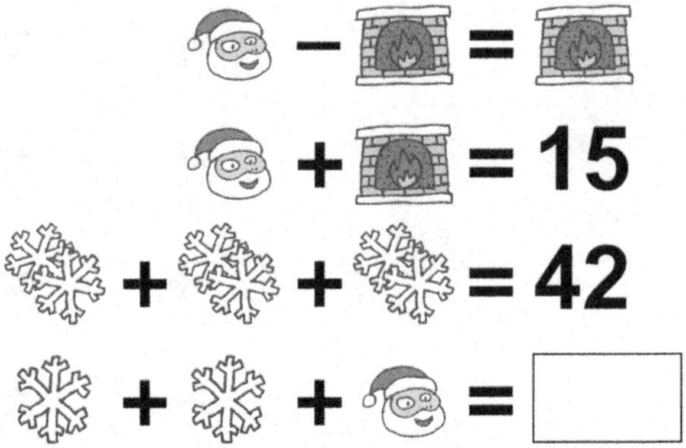

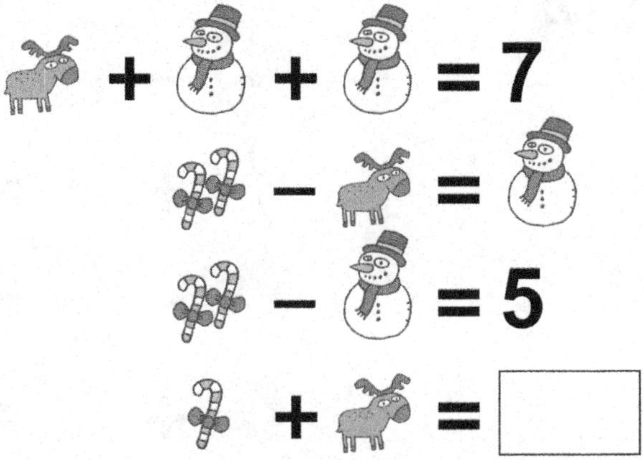

#118

$$\text{🎅❄} = 7$$

$$\text{❄} + \text{❄} = 4$$

$$\text{🎅} + \text{⛄} = 14$$

$$\text{⛄} - \text{🎅} + \text{❄} = \boxed{}$$

#119

$$\text{🔥} - \text{🎄} + \text{🎁} = \text{🎁}$$

$$\text{🎄} + \text{🎄} + \text{🎄} = 12$$

$$\text{🎁} + \text{🎁} = \text{🔥}$$

$$\text{🔥} - \text{🎁} + \text{🎄} = \boxed{}$$

#120

$$\text{❄} - \text{HO} = 5$$

$$\text{HO} + \text{HO} + \text{HO} = 18$$

$$\text{🍬} + \text{🍬} + \text{🍬} = \text{HO}$$

$$\text{❄} - \text{🍬} = \boxed{}$$

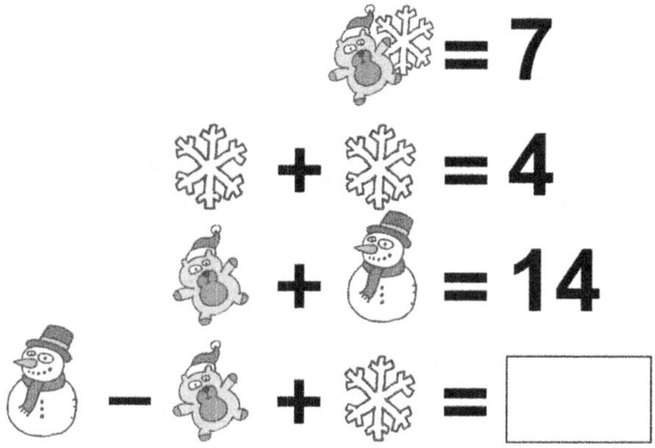

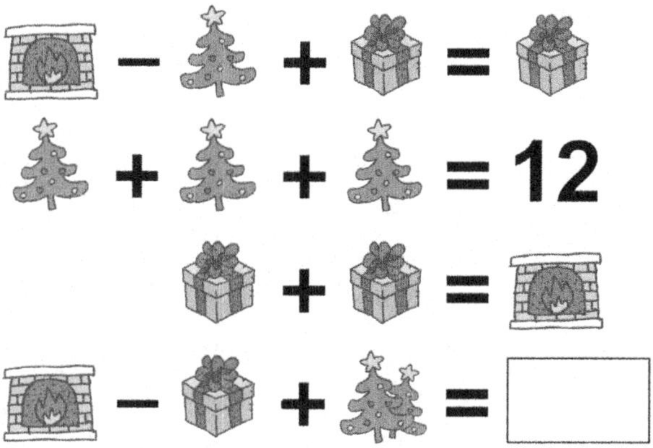

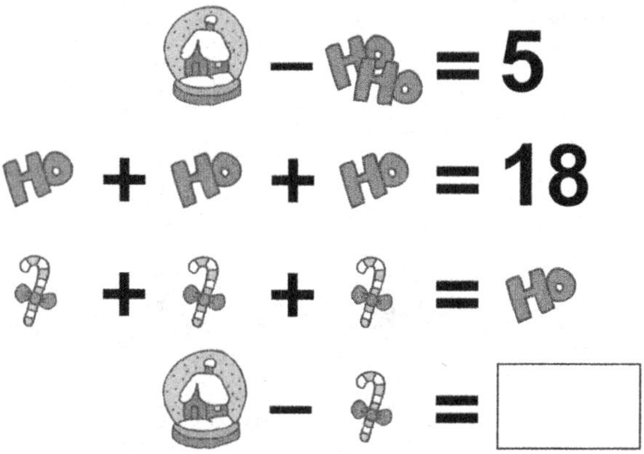

#121

#122

#123

🎄 − 🎁 = 16

🔥 + 🎁 = 14

🔥 + 🔥 + 🔥 = 18

🎄 − 🔥 =

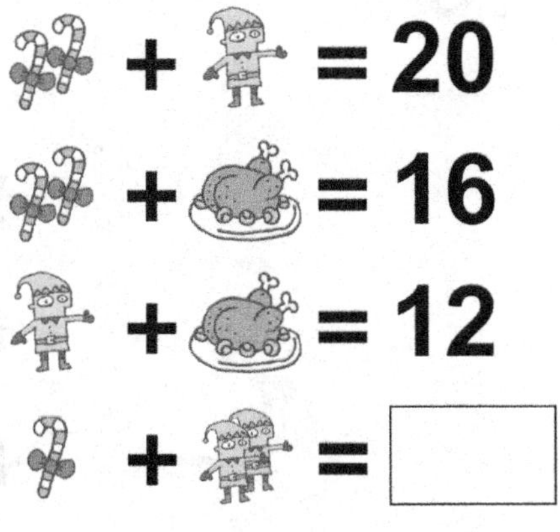

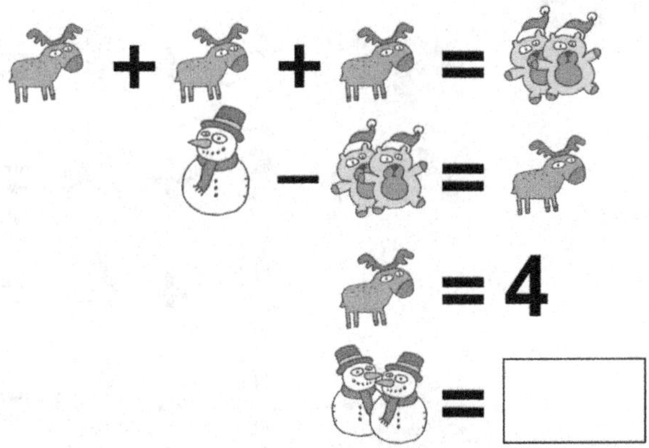

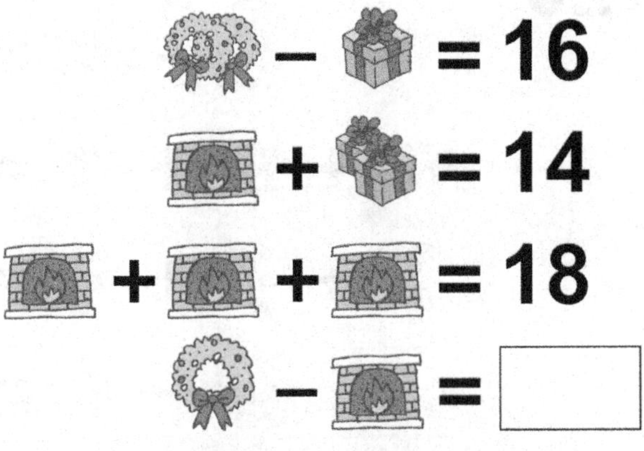

#124

 $-$  $= 6$

 $-$  $= 2$

 $-$  $=$

 $+$  $+$  $=$ ☐

#125

 $+$  $= 6$

 $-$  $= 4$

 $+$  $= 7$

 $+$  $+$  $=$ ☐

#126

 $+$  $= 13$

 $-$  $= 3$

 $+$  $+$  $=$

 $=$ ☐

#127

#128

$$\text{snowglobe} + \text{snowglobe} + \text{snowglobe} = 18$$

$$\text{fireplace} - \text{candycane} = \text{candycanes}$$

$$\text{candycanes} = \text{snowglobe}$$

$$\text{fireplace} = \boxed{}$$

#129

48

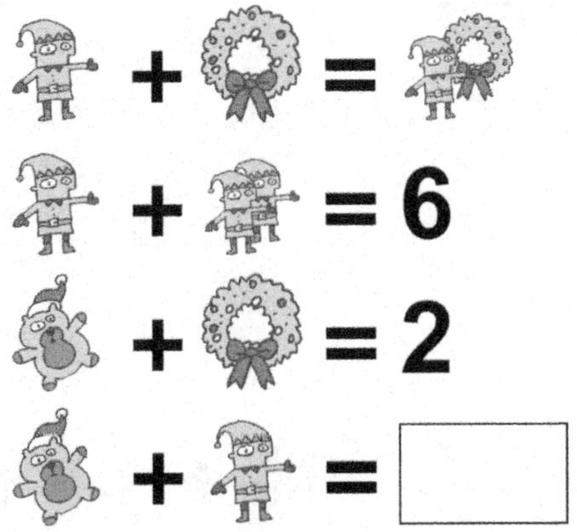

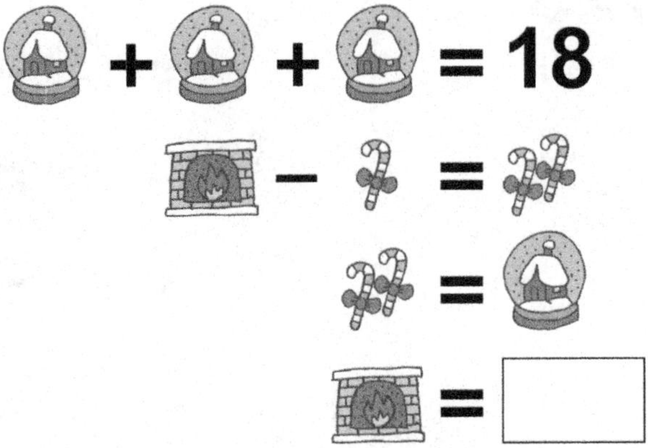

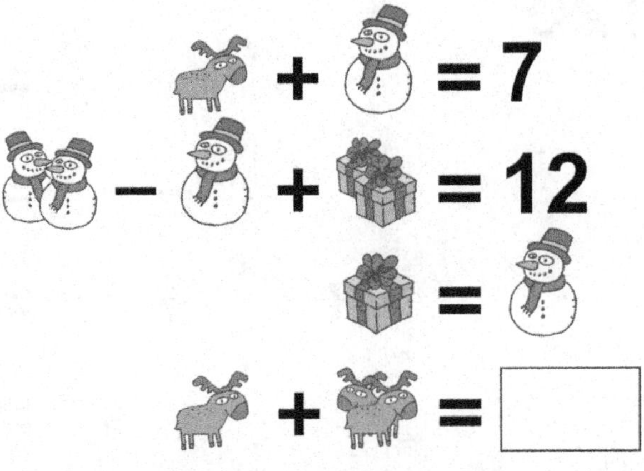

#130

$$\text{(elf)} - \text{(snowflake)} = 9$$
$$\text{(elf)} - \text{(snowflake)} - \text{(snowmen)} = 1$$
$$\text{(snowman)} + \text{(elf)} = 23$$
$$\text{(elf)} + \text{(snowflake)} = \boxed{}$$

#131

$$\text{(tree)} - \text{(hoho)} = 7$$
$$\text{(ho)} + \text{(candy cane)} = 13$$
$$\text{(candy cane)} + \text{(candy canes)} + \text{(candy cane)} = 36$$
$$\text{(tree)} + \text{(tree)} + \text{(tree)} = \boxed{}$$

#132

$$\text{(gift)} + \text{(wreath)} = 25$$
$$\text{(gift)} - \text{(wreath)} - \text{(santa)} = 2$$
$$\text{(santa)} + \text{(wreath)} = 14$$
$$\text{(wreath)} - \text{(santa)} = \boxed{}$$

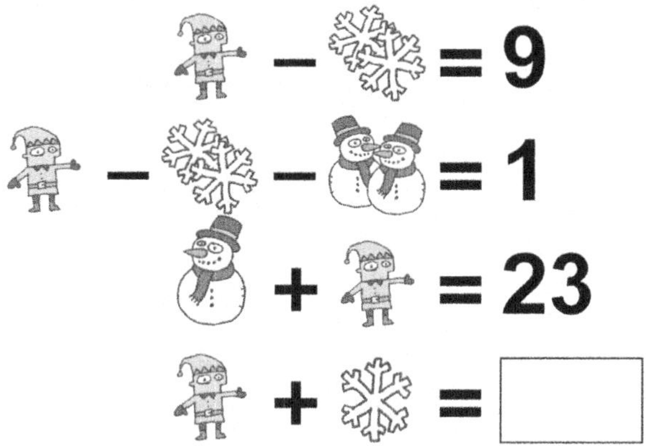

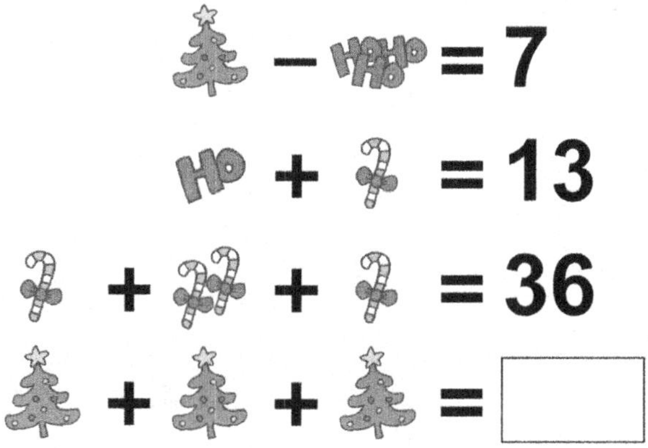

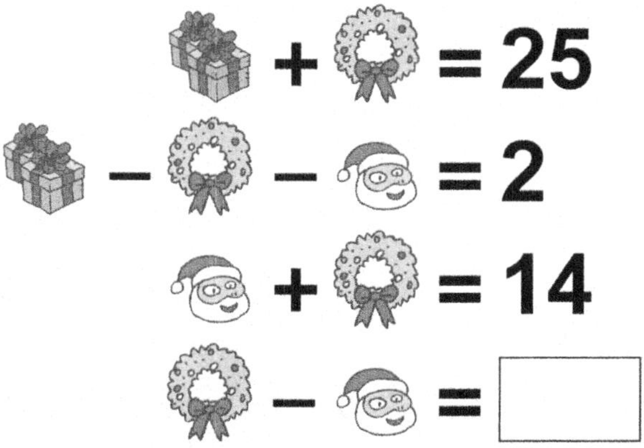

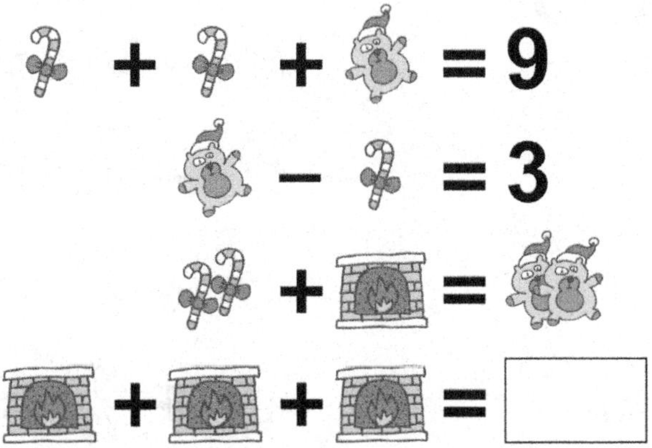

#133

$$\text{candy} + \text{candy} + \text{santa} = 9$$
$$\text{santa} - \text{candy} = 3$$
$$\text{candy candy} + \text{fireplace} = \text{santa santa}$$
$$\text{fireplace} + \text{fireplace} + \text{fireplace} = \boxed{}$$

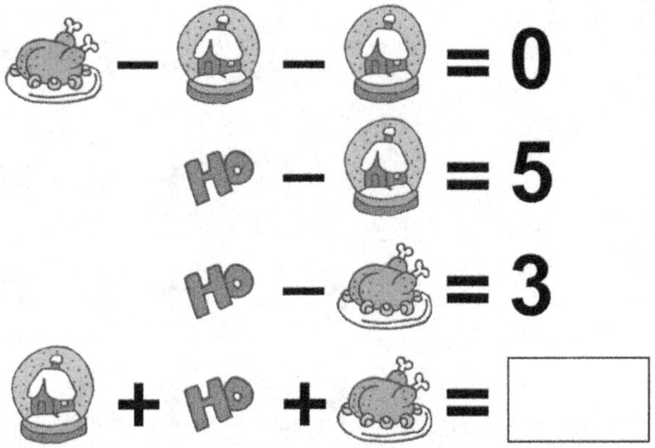

#134

$$\text{turkey} - \text{globe} - \text{globe} = 0$$
$$\text{Ho} - \text{globe} = 5$$
$$\text{Ho} - \text{turkey} = 3$$
$$\text{globe} + \text{Ho} + \text{turkey} = \boxed{}$$

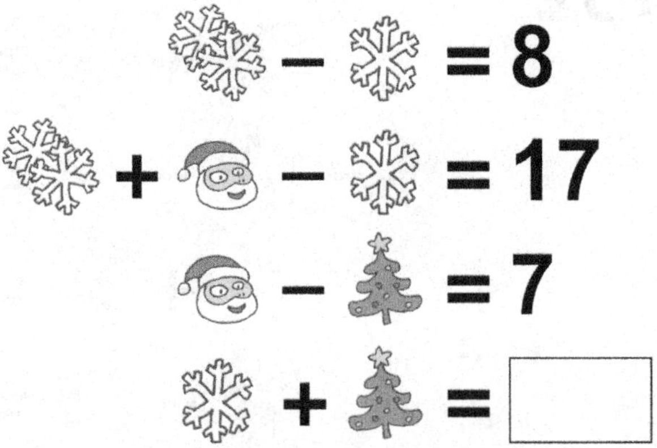

#135

$$\text{snowflakes} - \text{snowflake} = 8$$
$$\text{snowflakes} + \text{santa} - \text{snowflake} = 17$$
$$\text{santa} - \text{tree} = 7$$
$$\text{snowflake} + \text{tree} = \boxed{}$$

#136

❄ + ❄ + 🔮 = 10

🔮 + 🔮 = 8

HₐₒHₒ − ❄ − ❄ = 12

Hₒ − 🔮 = ☐

#137

⛄ − 🍗 = 5

⛄ + 🍬 + 🍬 = 19

🍬 − 🍗 = 1

⛄ + 🍗 − 🍬 = ☐

#138

🦌 − 🔥 + 🎄 = 7

🎄 + 🎄 = 4

🦌 + 🔥 = 9

🦌 + 🎄 = ☐

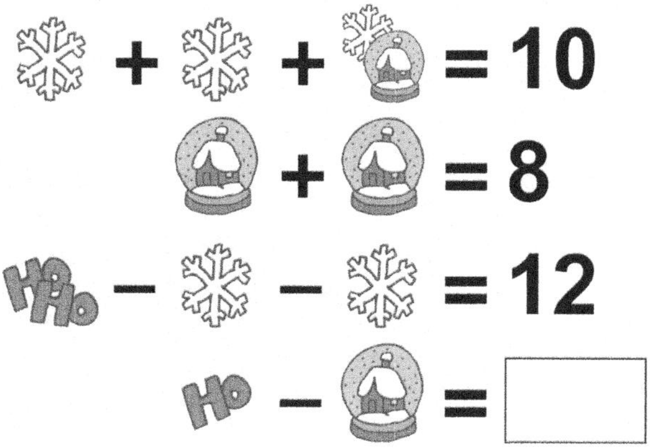

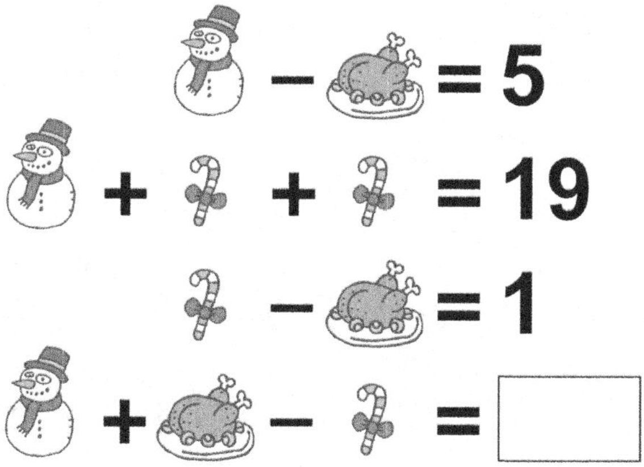

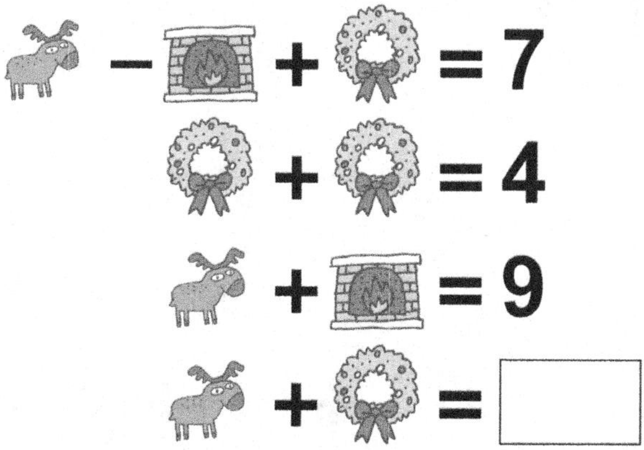

#139

🎁 − 🧝 = 4

🧝 + 🧝 = 100

🎁 + 🏠 + 🏠 = 45

🎁 − 🏠 = ☐

#140

⛄ + 🐷 = 10

🎄 + 🎄 + 🐷 = 13

🎄 + 🎄 + ⛄ = 17

⛄ − 🐷 + 🎄 = ☐

#141

🔮 + 🦌 = 12

❄ − 🦌 + 🦌 = 5

❄ − 🔮 = 3

❄ + ❄ + 🦌 = ☐

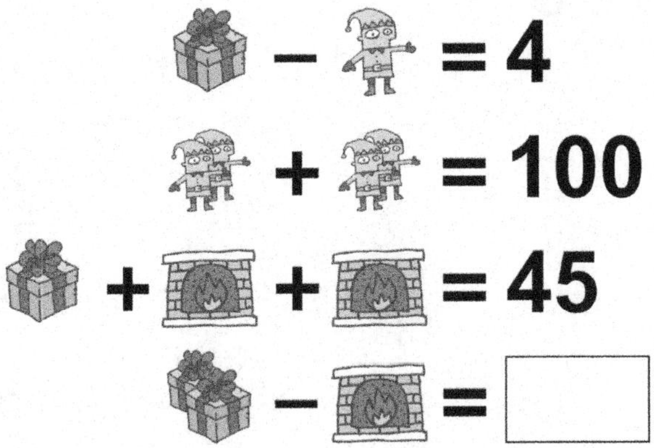

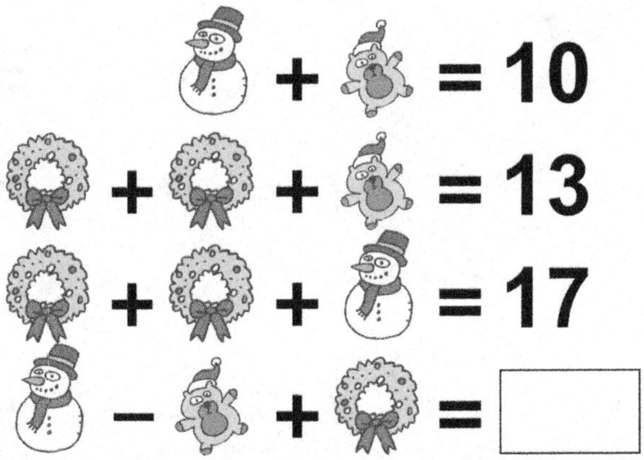

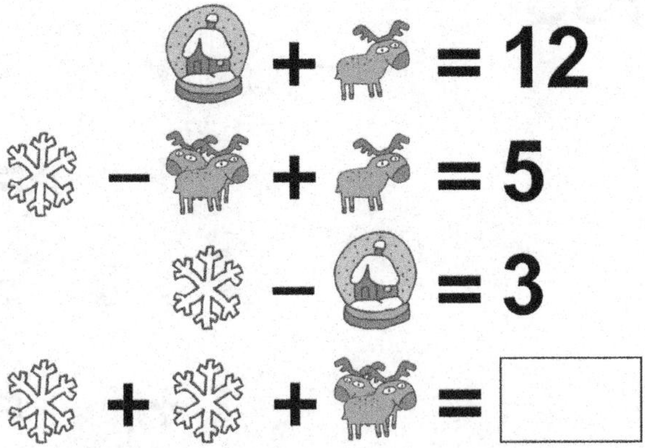

#142

$$\text{❄} + \text{🎁} = 9$$

$$\text{Ho🎁} + \text{🎁} = 10$$

$$\text{❄} + \text{Ho} + \text{🎁} = 10$$

$$\text{HoHoHo} - \text{❄} - \text{🎁} = \boxed{}$$

#143

$$\text{🐷🐷} = \text{🔮}$$

$$\text{⛄} - \text{🐷} = \text{🔮}$$

$$\text{⛄} + \text{🐷} = 12$$

$$\text{⛄} + \text{🔮} = \boxed{}$$

#144

$$\text{🔥} + \text{🧝} + \text{🎅} = \text{🎅🧝}$$

$$\text{🎅} - \text{🧝} - \text{🧝} = \text{🧝}$$

$$\text{🎅} = 90$$

$$\text{🔥} = \boxed{}$$

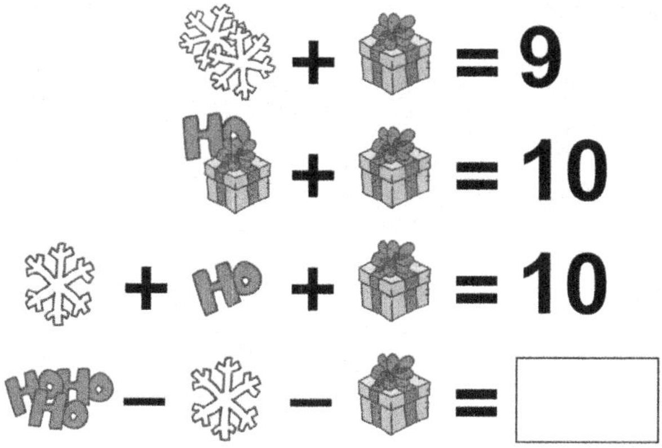

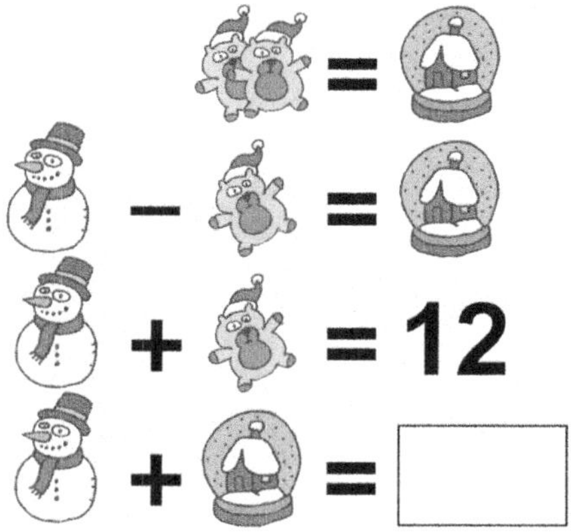

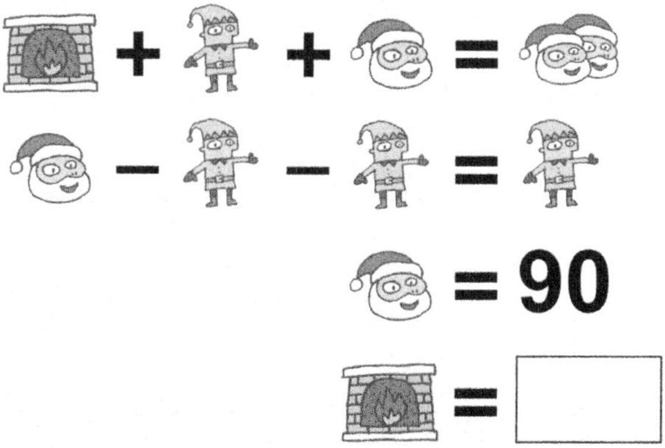

#145

🍬 − ☃ = 2

☃ + 🍗 + 🍗 = 🍬

☃ − 🍗 − 🍗 = 3

🍬 − ☃ + 🍬 = ▭

#146

🧝🎄 − 🔥 = 4

🎄 + 🎄 + 🎄 = 9

🔥 + 🧝 = 13

🧝 + 🧝 − 🔥 = ▭

#147

🎅 − HO = 🎄

🎅 + 🎅 − HO = 17

🎅 − HO = 2

🎄 − 🎅 + HO = ▭

54

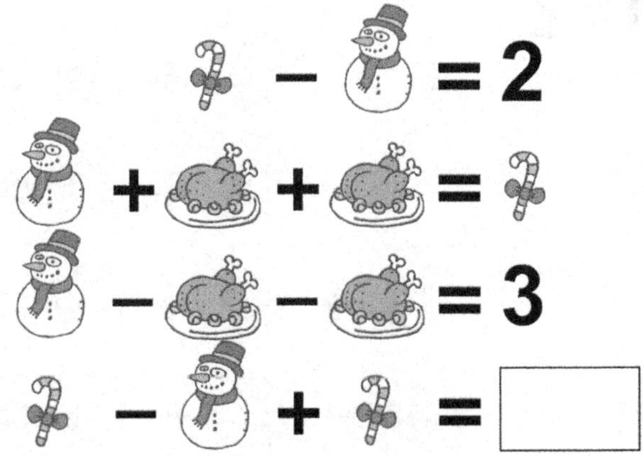

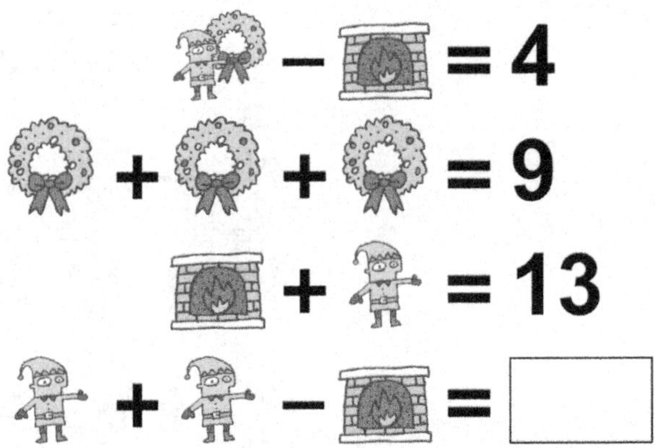

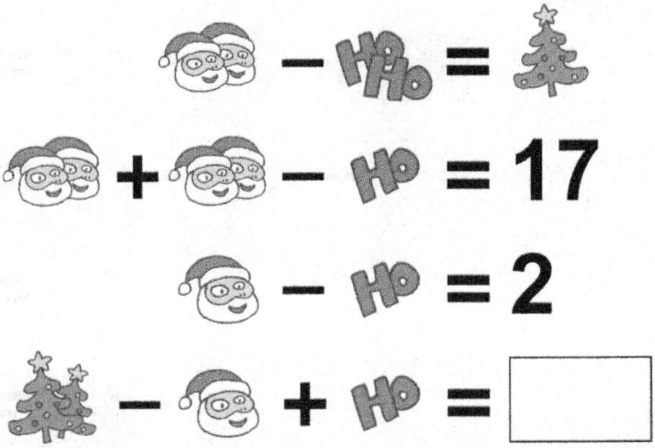

#148

🧝 + 🧝 + 🧝 = 15

🎁 − 🧝 = ❄️

❄️ − 🧝 = 2

🎁 + ❄️ + 🧝 = ☐

#149

🦃 + 🦃 − 🎀 = 0

🔥 − 🦃 − 🦃 = 2

🎀 + 🔥 = 10

🎀 − 🔥 = ☐

#150

🎄 − 🔮 = 🍬

🔮 + 🔮 + 🎄 = 35

🍬 + 🍬 = 10

🍬 + 🎄 + 🔮 = ☐

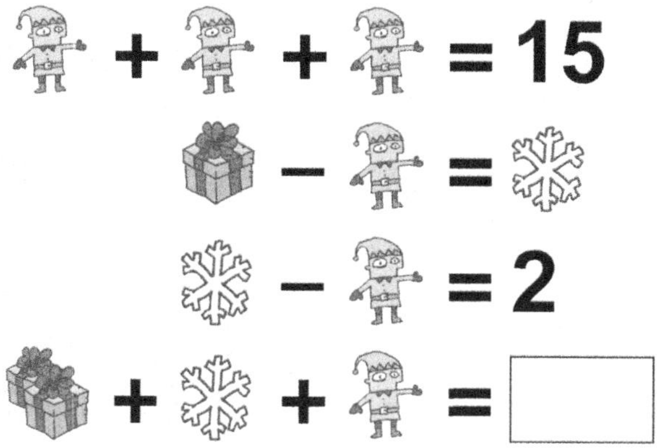

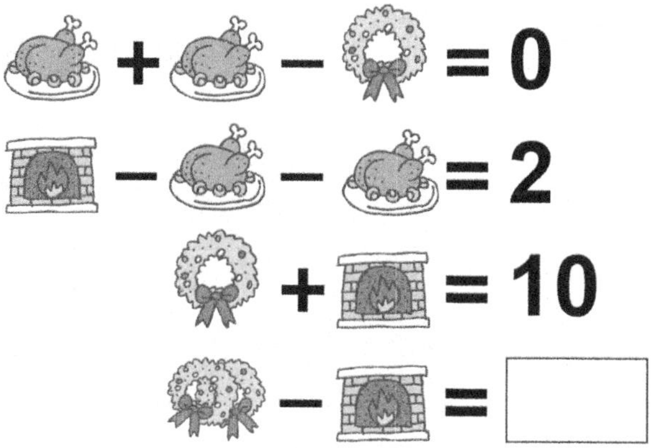

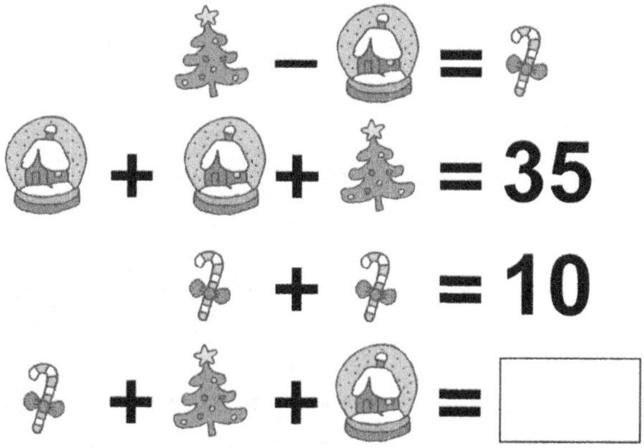

#151

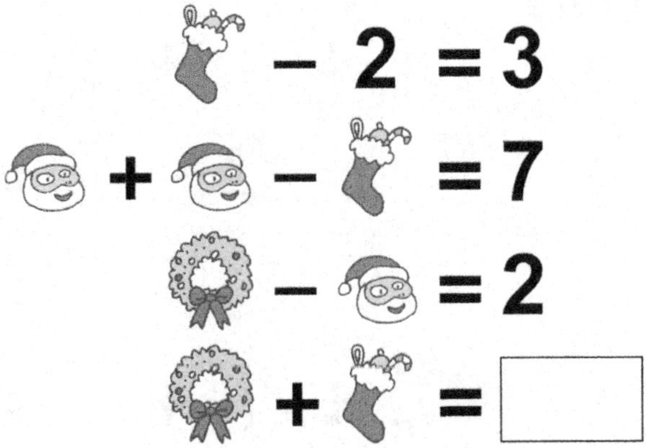

$$\text{stocking} - 2 = 3$$
$$\text{santa} + \text{santa} - \text{stocking} = 7$$
$$\text{wreath} - \text{santa} = 2$$
$$\text{wreath} + \text{stocking} = \boxed{}$$

#152

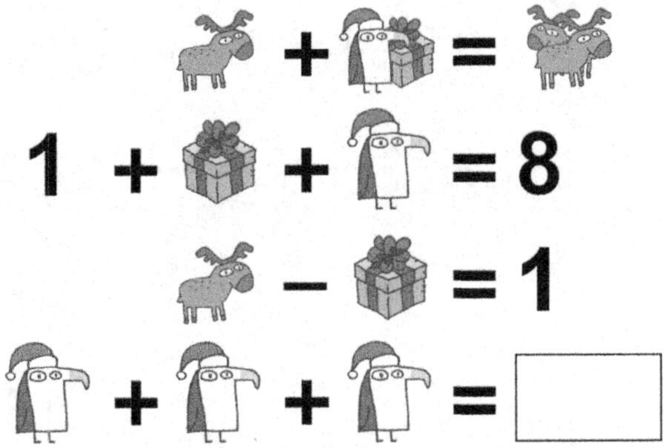

$$\text{moose} + \text{gift} = \text{two moose}$$
$$1 + \text{gift} + \text{elf} = 8$$
$$\text{moose} - \text{gift} = 1$$
$$\text{elf} + \text{elf} + \text{elf} = \boxed{}$$

#153

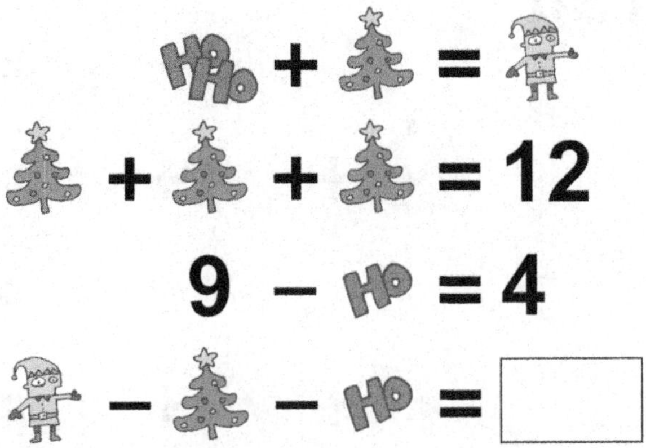

$$\text{HoHo} + \text{tree} = \text{elf}$$
$$\text{tree} + \text{tree} + \text{tree} = 12$$
$$9 - \text{Ho} = 4$$
$$\text{elf} - \text{tree} - \text{Ho} = \boxed{}$$

#154

🎁 − 🧝 = **10**

❄ + ❄ = 🎁

❄ − 🧝 = **4**

🎁 + ❄ + 🧝 = ⬜

#155

�de☃ + 🎍 = 🇧🇧

🦌 + 🦌 = ☃

🎍 + 🦌 = **21**

🎍 − 🦌 − 🦌 = ⬜

#156

😊 + 😊 + 😊 = 🎅

HOHO + HO + HOHO = 🎅

🎅 − HOHO − HO = **36**

🎅 − 😊 − 😊 = ⬜

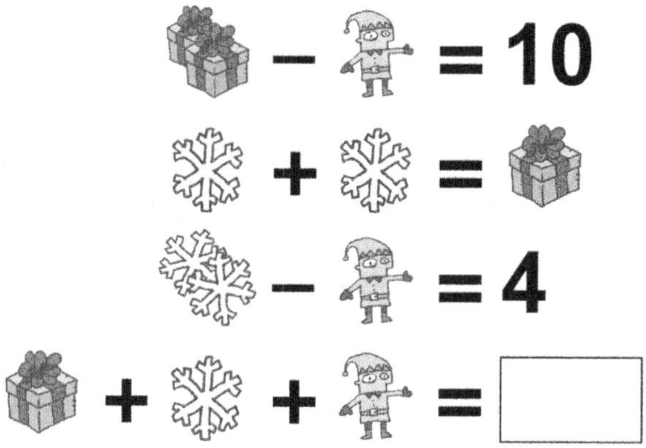

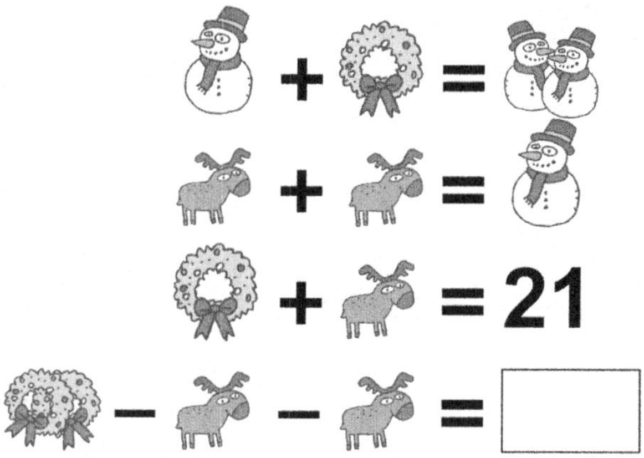

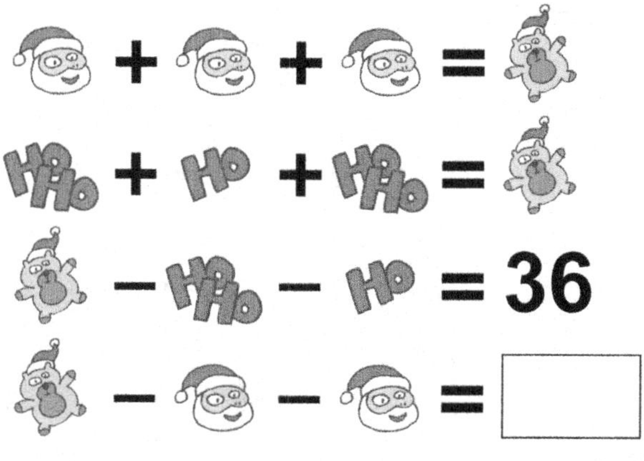

#157

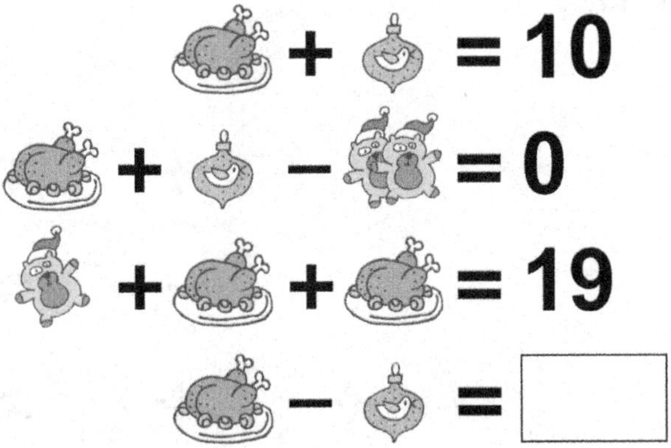

#158

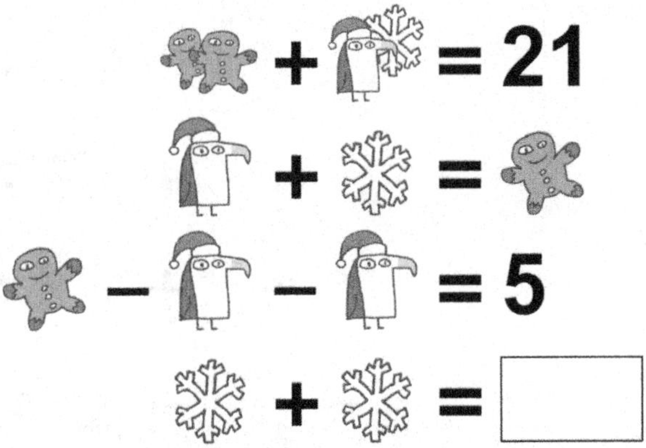

#159

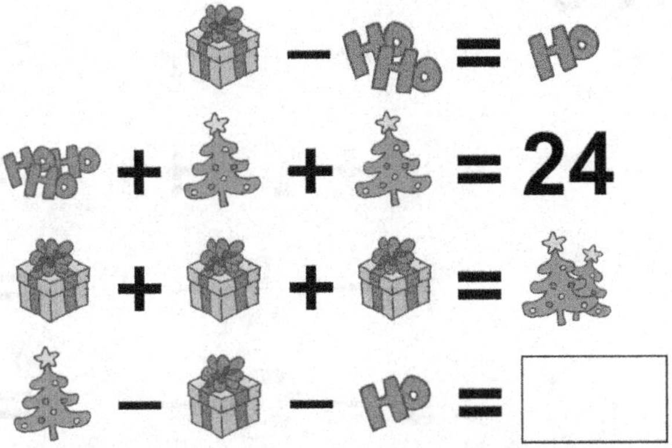

#160

🍬 − ❄️ = 7

🍬 + ❄️ = 13

🎄 − 🍬 = 1

🎄 + ❄️ + ❄️ = ☐

#161

🎁 + 🔥 = 9

🎁 + 🔥 = 🔮

🔮 + 🔮 = 16

🔮 − 🎁 + 🔥 = ☐

#162

🦌 + 🦌 + 🦌 = ⛄

🦌 + 🦌 + 🦌 = 🍗

⛄ − 🍗 − 🦌 = 12

🍗 − 🦌 − 🦌 = ☐

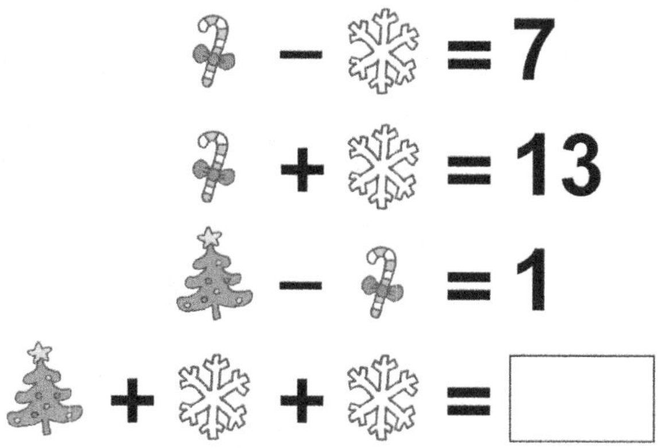

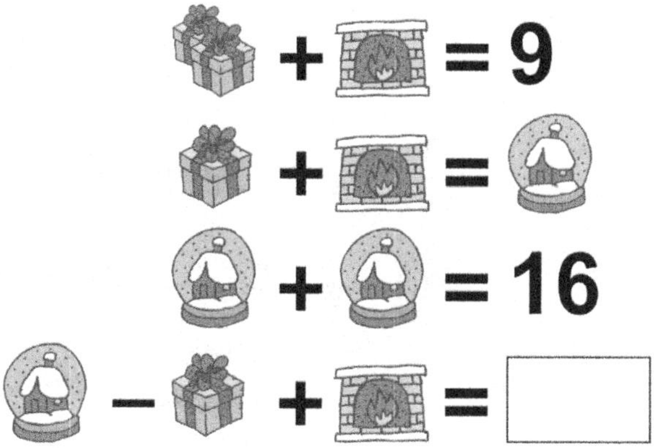

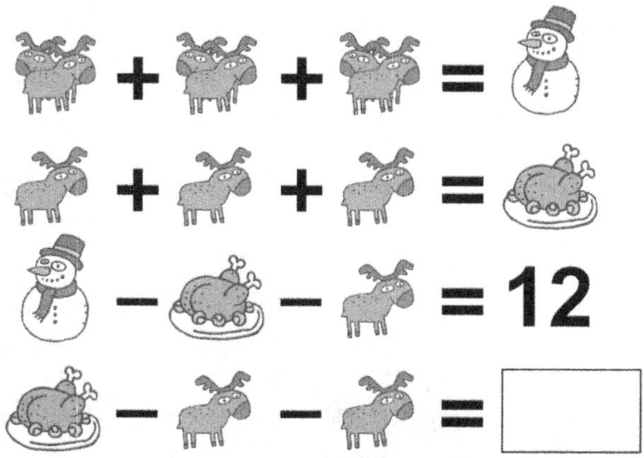

#163

 − = 2

 + 3 + = 13

 + = 8

 − = ☐

#164

 − = 1

 + + 4 = 15

 + − = 9

 − = ☐

#165

 + =

 − + = 15

 − = 1

 + + = ☐

#166

#167

#168

61

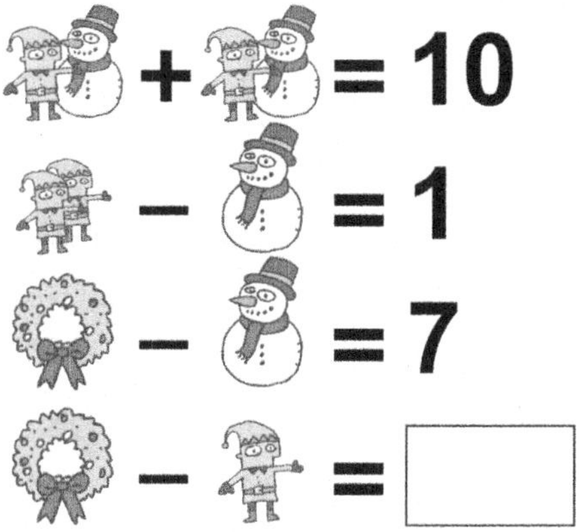

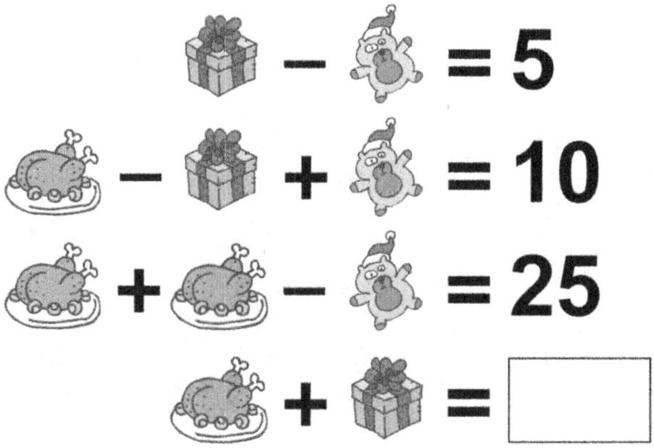

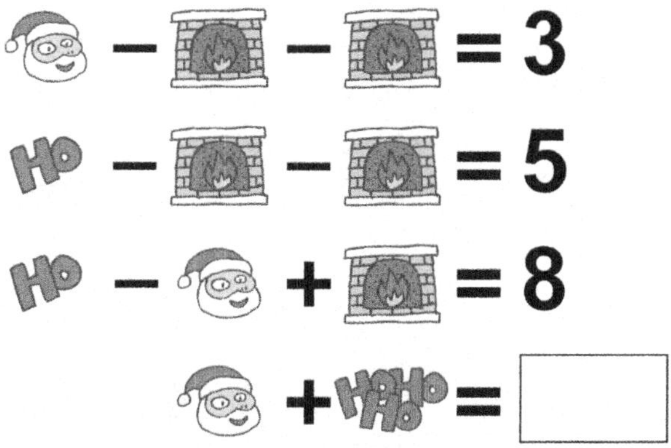

#169

$$\text{(snow globe)} + \text{(snowflake)} = 14$$
$$\text{(snow globe)} - \text{(elf)} = 3$$
$$\text{(snowflake)} - \text{(elf)} = 5$$
$$\text{(elf)} + \text{(elf)} = \boxed{}$$

#170

$$\text{HoHo} + \text{HoHo} = 30$$
$$\text{Ho} + \text{(tree)} + \text{(candy cane)} = 18$$
$$\text{Ho} - \text{(tree)} + \text{(candy cane)} = 12$$
$$\text{(tree)} + \text{(candy cane)} + \text{HoHo} = \boxed{}$$

#171

$$\text{(moose)} - \text{(fireplace)} + \text{(turkey)} = \text{(moose)}$$
$$\text{(turkey)} + \text{(fireplace)} + \text{(turkey)} = 18$$
$$\text{(moose)} - \text{(turkey)} = 4$$
$$\text{(moose)} + \text{(moose)} = \boxed{}$$

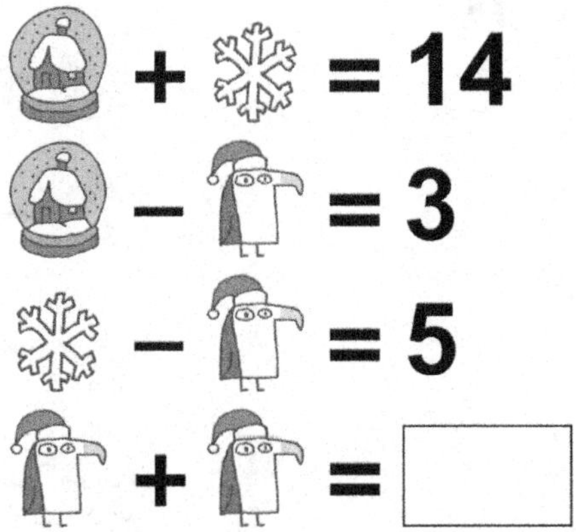

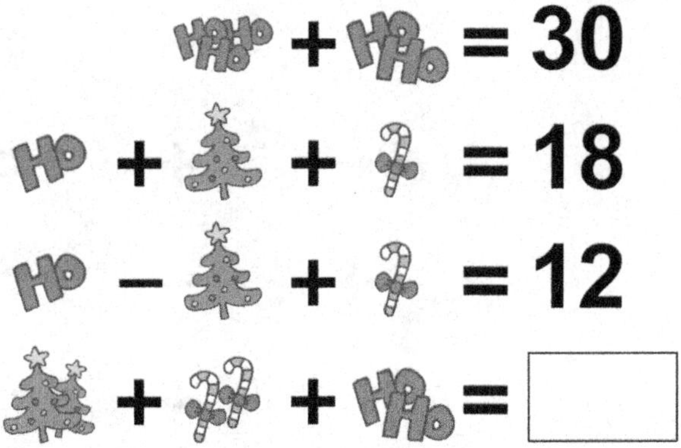

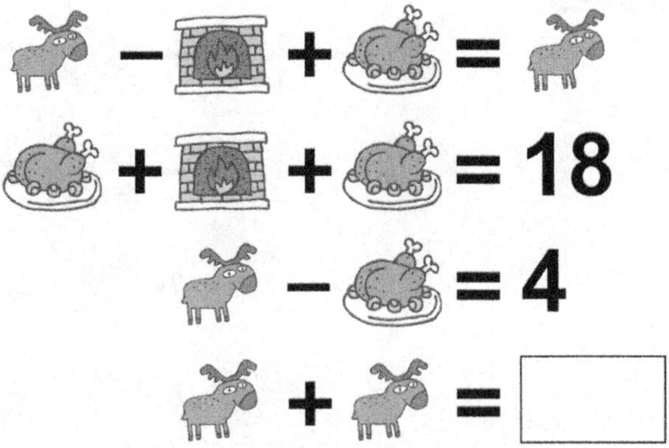

#172

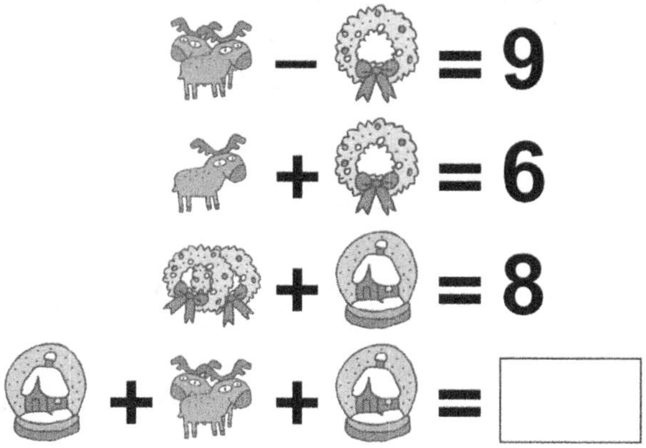

$$\text{(reindeer)} - \text{(wreath)} = 9$$
$$\text{(moose)} + \text{(wreath)} = 6$$
$$\text{(wreath)} + \text{(snow globe)} = 8$$
$$\text{(snow globe)} + \text{(reindeer)} + \text{(snow globe)} = \boxed{}$$

#173

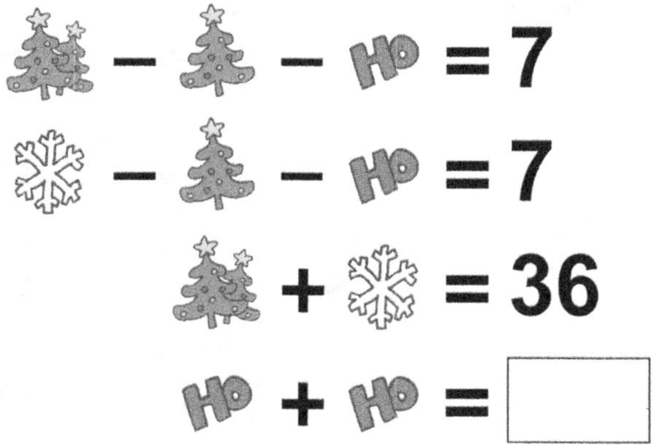

$$\text{(tree)} - \text{(tree)} - \text{(HO)} = 7$$
$$\text{(snowflake)} - \text{(tree)} - \text{(HO)} = 7$$
$$\text{(tree)} + \text{(snowflake)} = 36$$
$$\text{(HO)} + \text{(HO)} = \boxed{}$$

#174

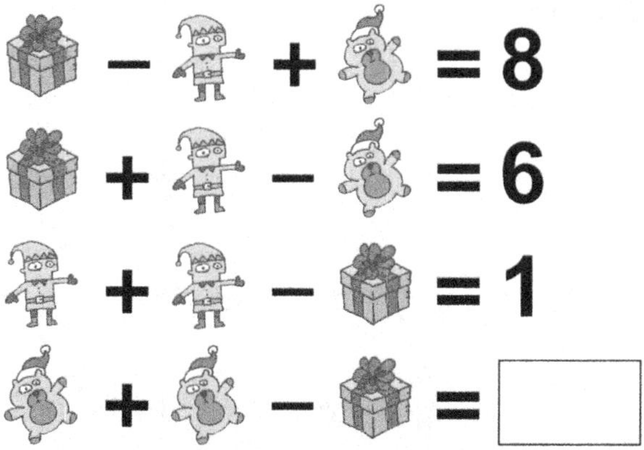

$$\text{(gift)} - \text{(elf)} + \text{(santa)} = 8$$
$$\text{(gift)} + \text{(elf)} - \text{(santa)} = 6$$
$$\text{(elf)} + \text{(elf)} - \text{(gift)} = 1$$
$$\text{(santa)} + \text{(santa)} - \text{(gift)} = \boxed{}$$

#175

$12 - \text{🍪} = \text{🎅}$

 $= 12$

$\text{🎅} - \text{🎅} = 5$

$\text{🎅} + \text{🍪} = \boxed{}$

#176

$\text{🎄} + \text{👼} = 22$

$\text{👼} - 5 = 5$

$\text{🎄} + \text{🔔} - \text{👼} = 17$

$\text{🔔} - \text{🎄} = \boxed{}$

#177

$\text{🧦} - \text{🎁} = 1$

$\text{🐧} + \text{🐧} = \text{🧦}$

$\text{🐧} - \text{🎁} + \text{🧦} = 9$

$\text{🎁} + \text{🎁} + \text{🎁} = \boxed{}$

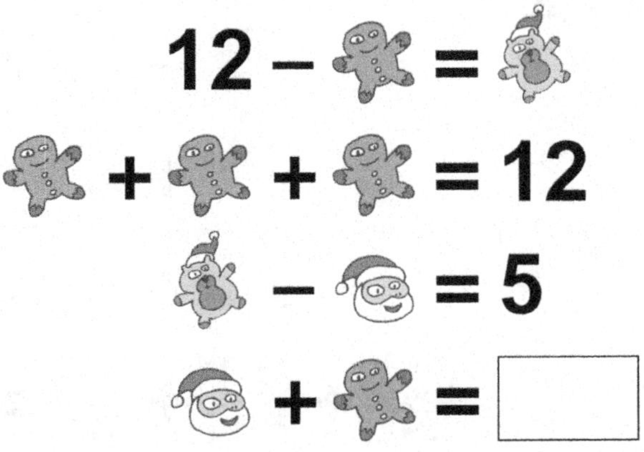

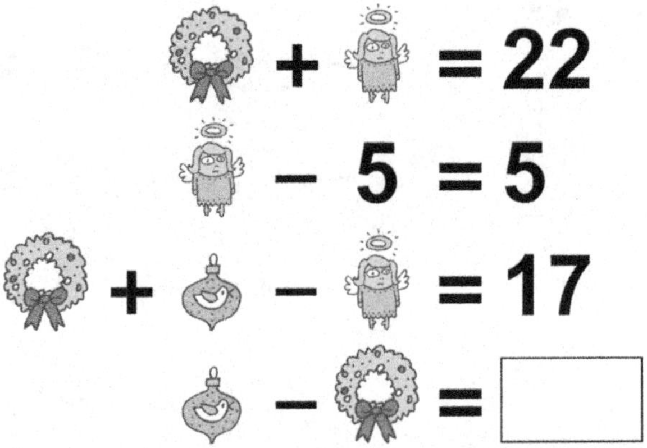

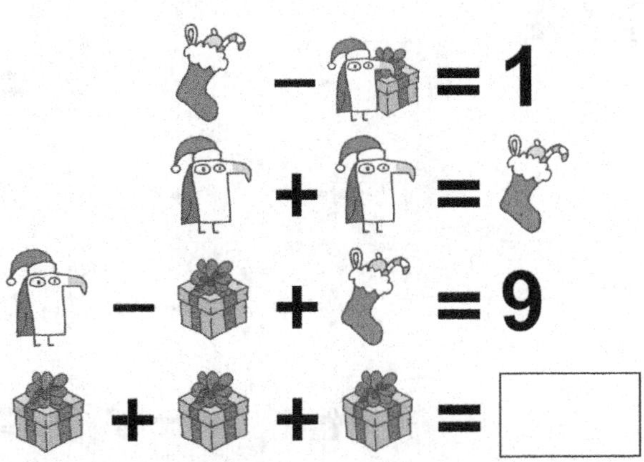

#178

🍬 + 🧝 = **20**

🔥 + 🧝 = **11**

🔥 + 🍬 = **15**

🧝 + 🔥 + 🍬 = ☐

#179

🎄 + 🎄 + ❄ = **20**

🐷 − ❄ = **12**

🎄 + 🎄 + 🐷 = **20**

🐷 − ❄ − 🎄 = ☐

#180

🎅 − 🍗 = ☃☃

☃ + ☃ + ☃ = 🎅

☃ + 🎅 + 🍗 = **30**

🎅 + 🎅 + 🎅 = ☐

65

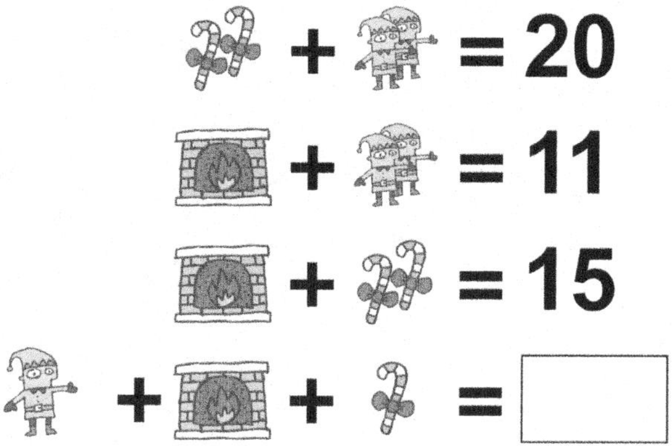

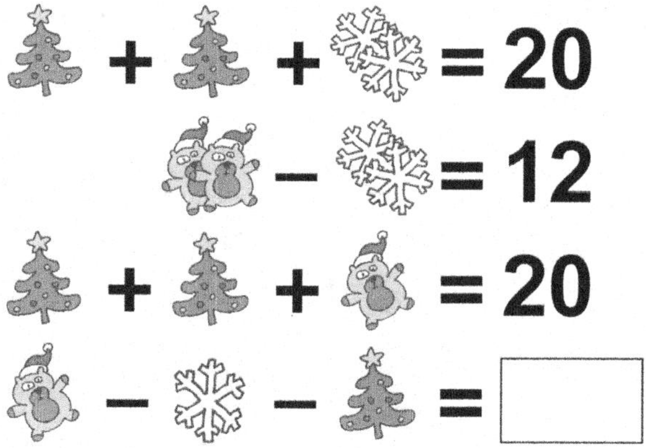

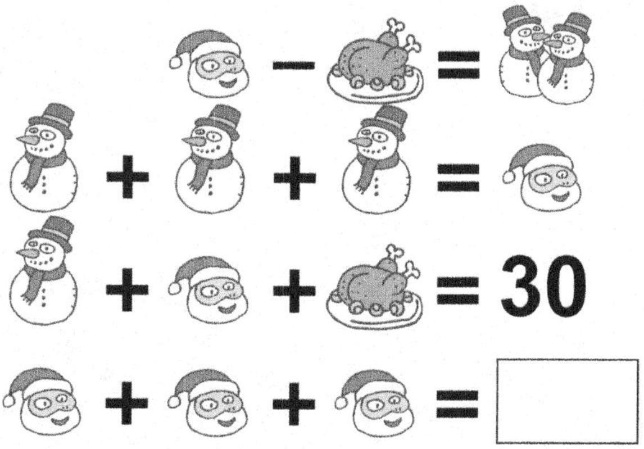

#181

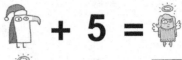

#182

#183

#184

$$\text{(snow globe)} - \text{(snowman)} = 5$$
$$\text{(snowflake)} + \text{(snowflake)} - \text{(snowman)} = 3$$
$$\text{(snow globe)} + \text{(snow globe)} = 12$$
$$\text{(snowflake)} + \text{(snowman)} + \text{(snowman)} = \boxed{}$$

#185

$$\text{(fireplace)} + \text{(fireplace)} = \text{(reindeer)}$$
$$\text{(fireplace)} + \text{(fireplace)} = \text{(tree)}$$
$$\text{(tree)} - \text{(reindeer)} = 16$$
$$\text{(tree)} + \text{(reindeer)} + \text{(fireplace)} = \boxed{}$$

#186

$$\text{(wreath)} + \text{(elf)} = 16$$
$$\text{(wreath)} - \text{(elf)} = 5$$
$$\text{(elf)} - \text{(gift)} + \text{(elf)} = 3$$
$$\text{(wreath)} - \text{(gift)} = \boxed{}$$

#187

$$\text{reindeer} - \text{elf} = 3$$
$$\text{reindeer} + \text{reindeer} + \text{reindeer} = 42$$
$$\text{reindeer} - \text{stocking} = 4$$
$$\text{stocking} + \text{elf} + \text{stocking} = \boxed{}$$

#188

$$\text{tree} + \text{tree} + \text{santa} = 16$$
$$\text{ornament} - \text{tree} - \text{tree} = 0$$
$$\text{ornament} - \text{santa} - \text{santa} = \text{tree}$$
$$\text{tree} - \text{santa} - \text{santa} = \boxed{}$$

#189

$$\text{snowflake} + \text{snowflake} = \text{wreath}$$
$$\text{wreath} - \text{snowflake} = 9$$
$$\text{gingerbread} + \text{snowflake} = 16$$
$$\text{gingerbread} + \text{wreath} = \boxed{}$$

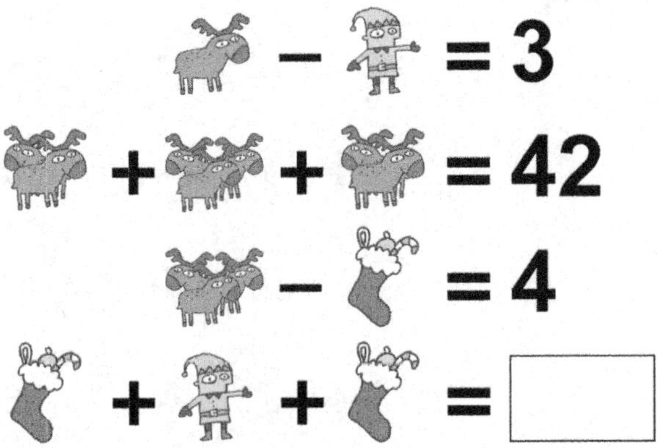

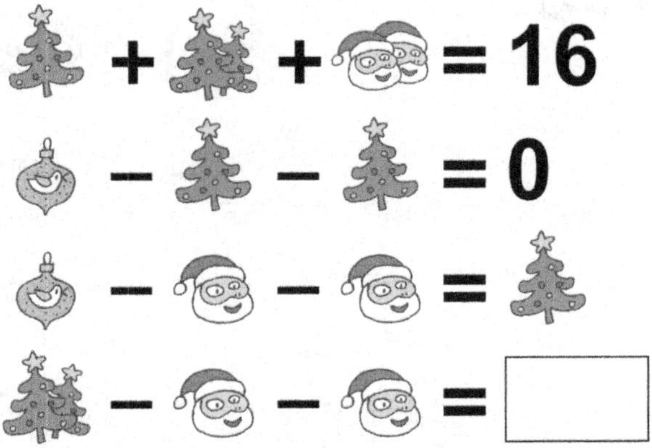

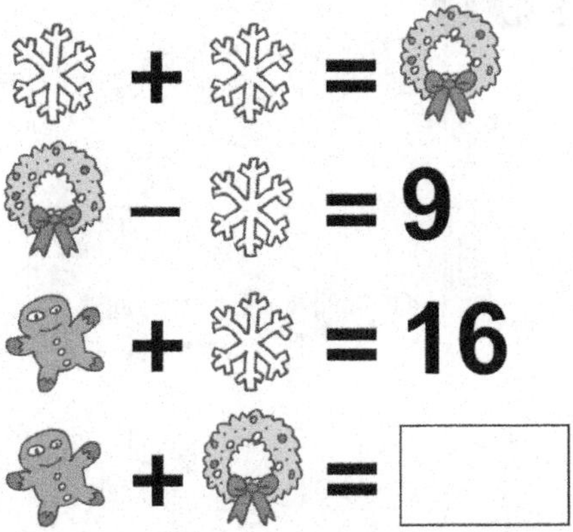

#190

$$\text{🎅} + \text{🦌} = 23$$

$$\text{🦌} - \text{🎄} = 1$$

$$\text{⛄} - 7 - \text{🎄} = 1$$

$$\text{⛄} - \text{🦌} + \text{🎄} = \boxed{}$$

#191

$$\text{❄} + \text{HoHo} + \text{❄} = 45$$

$$\text{HoHo} + \text{❄} + \text{🎅} = 50$$

$$\text{🎅} + \text{HoHo} + \text{🎅} = 45$$

$$\text{Ho} + \text{🎅} + \text{🎅} = \boxed{}$$

#192

$$\text{⛄} + \text{🎁} = 14$$

$$\text{🎁} + \text{🍗} = \text{⛄}$$

$$\text{🍗} + \text{🍗} = \text{🎁}$$

$$\text{⛄⛄} + \text{🎁} = \boxed{}$$

69

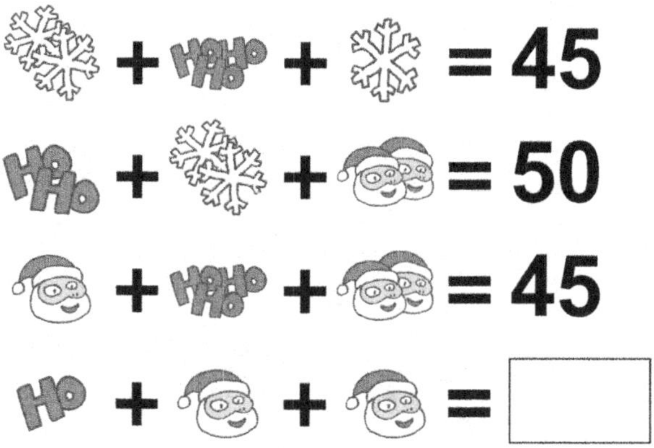

#193

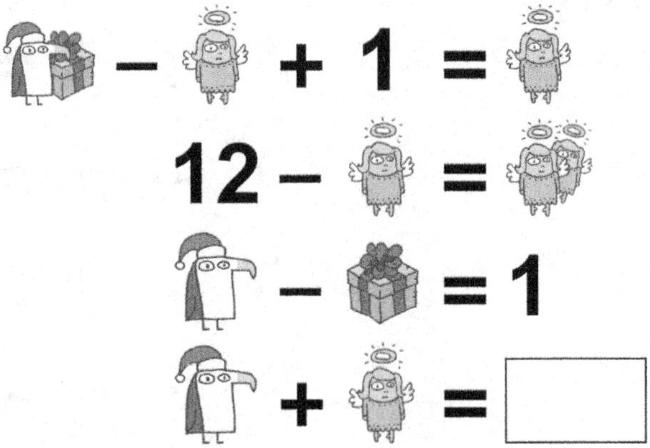

#194

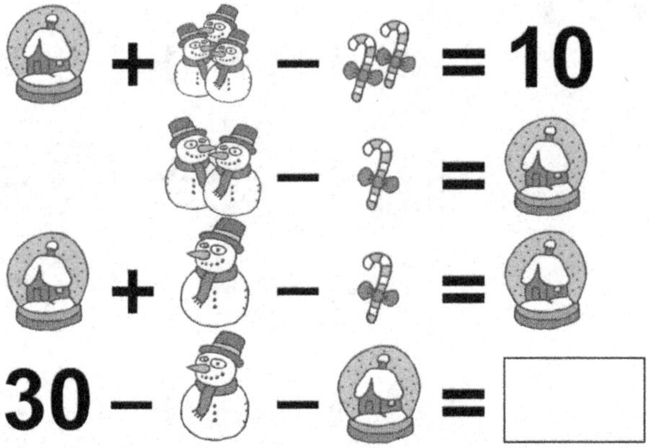

#195

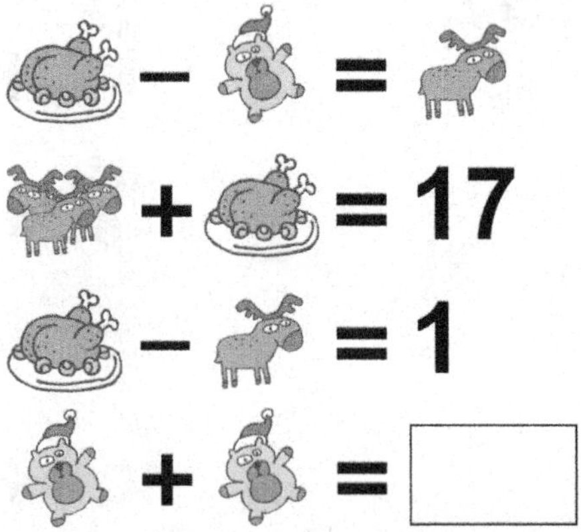

70

#196

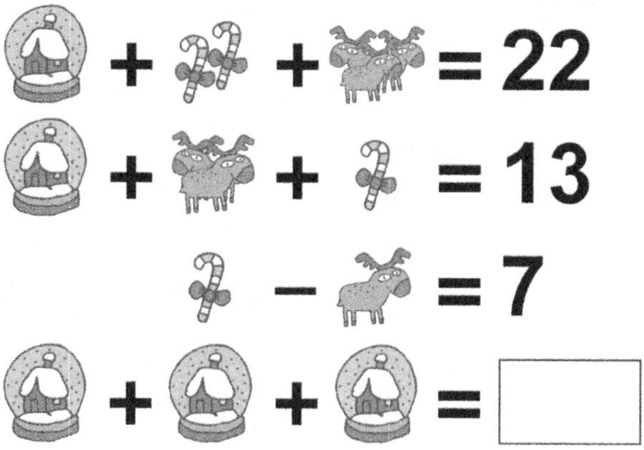 = 22

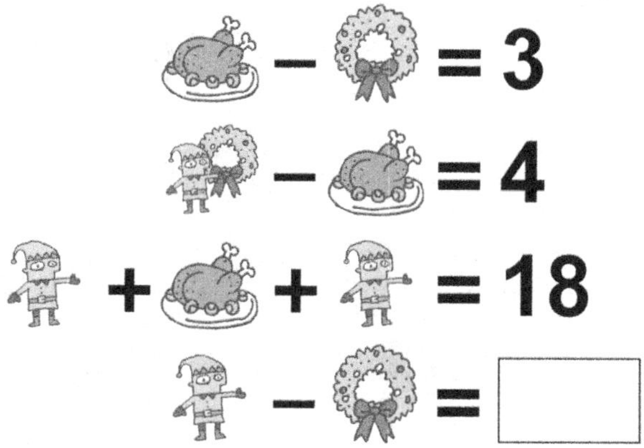 = 13

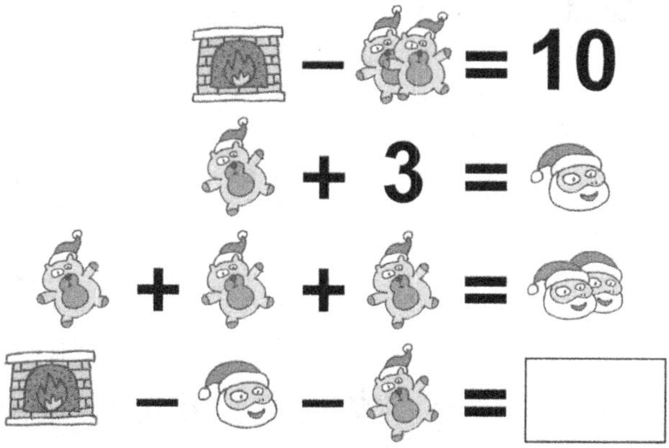 = 7

🔮 + 🔮 + 🔮 =

#197

🍗 − 🎀 = 3

🎀 − 🍗 = 4

🧝 + 🍗 + 🧝 = 18

🧝 − 🎀 =

#198

🔥 − 👥 = 10

🎅 + 3 = 😊

🎅 + 🎅 + 🎅 = 😊😊

🔥 − 😊 − 🎅 =

71

#199

12 − 🧑 + 👼 = 🔥

7 + 🧑 = 13

🔥 − 👼 = ⬜

#200

❄ + ❄ = 🧝

🎁 + ❄ = 🧝

🧝 − 🎁 = 4

20 − 🧝 + 🎁 = ⬜

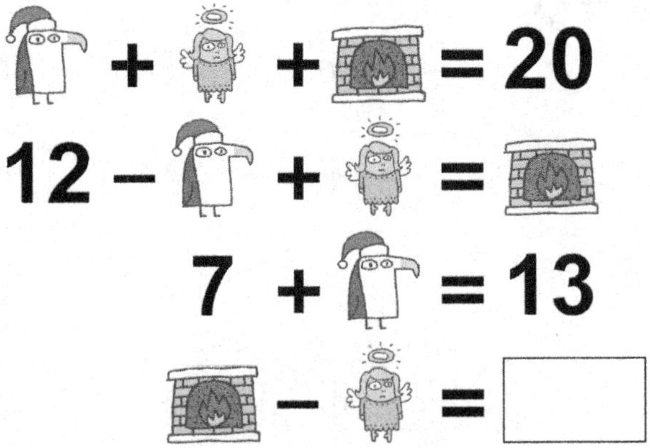

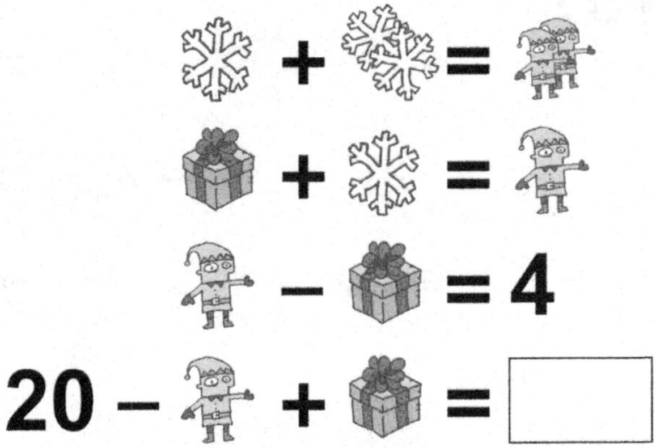

#201

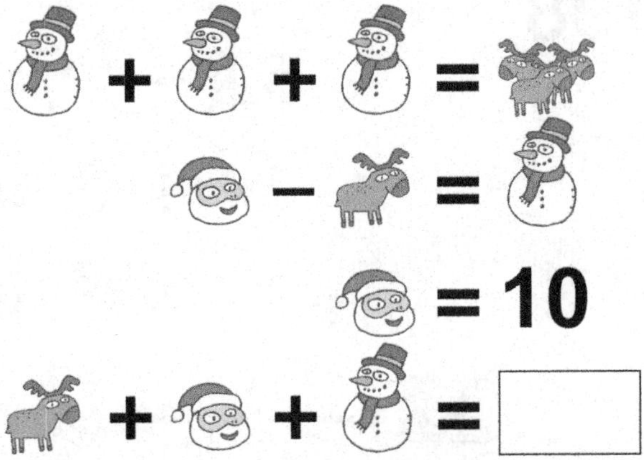

#202

$$\text{stocking} - \text{gingerbread}\times 2 = 4$$
$$\text{angel} + \text{angel}\times 2 + \text{angel} = \text{stocking}$$
$$\text{angel} + \text{stocking} = 21$$
$$\text{gingerbread} + \text{gingerbread} + \text{gingerbread} = \boxed{}$$

#203

$$\text{reindeer} + \text{HO} = \text{snowflake}$$
$$\text{HOHO} - \text{reindeer} = \text{snowflake}$$
$$\text{reindeer} + \text{reindeer} + \text{reindeer} = 3$$
$$\text{snowflake} + \text{snowflake} = \boxed{}$$

#204

$$\text{ornament} + \text{gift} = 15$$
$$\text{ornament} - 3 + \text{santa+gift} = 17$$
$$\text{santa} - \text{gift} = 1$$
$$\text{ornament} - \text{santa+gift} + \text{santa} = \boxed{}$$

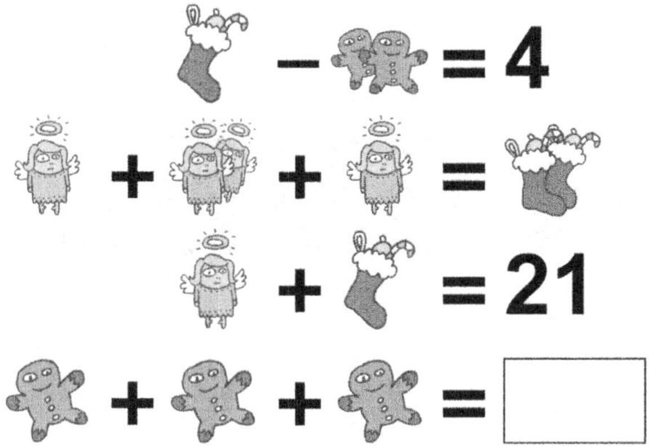

LEVEL 3

This final set of puzzles is very tricky. You'll need to think outside the box to solve them. Here are some tips to help you get to the answers.

1. Did you know? You can combine two equations to get one bigger equation. For example:

EQUATION 1:   🎅 + ⛄ = 3

EQUATION 2:   🎅 − ⛄ = 1

BECOMES...   🎅 + ⛄ + 🎅 − ⛄ = 3 + 1

                                              🎅🎅 = 4

                                              🎅 = 2

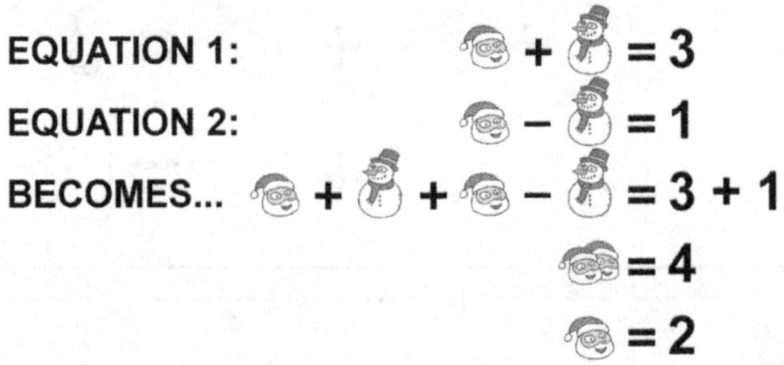

2. If a symbol seems like it can have more than one value, assume a value for it, then see if everything else works.

Alright, let's do this!

#205

 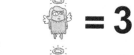 = 3

 x  = 6

  − = 1

 + + =

#206

 = 2

 + = 7

 x  = 15

 − =

#207

 +  = 10

 + = 3

 − = 3

 + =

#208

🎄 + 🎀 = 13

🎅 − 🎀 = 3

🎅 + 🎅 + 🎅 = 🎄

🎅 = ☐

#209

📝 + 🍪 = 13

👻 + 📝 = 12

🍪 + 👻 = 11

📝 X 📝 = ☐

#210

🍰 + 🌷 = 30

🧦 − 🌷 = 4

🌷 + 🧦 = 18

🍰 + 🧦 + 🍰 = ☐

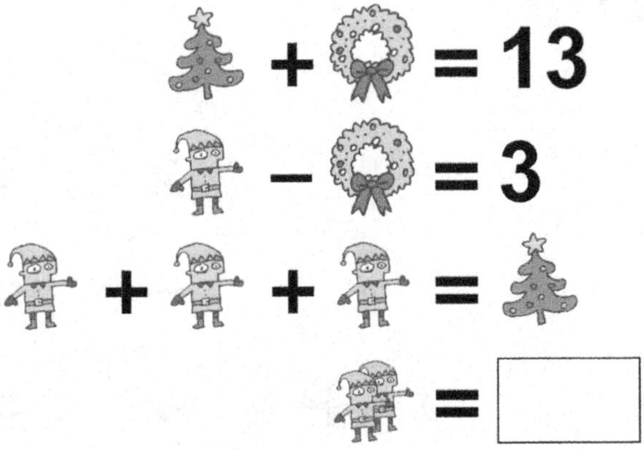

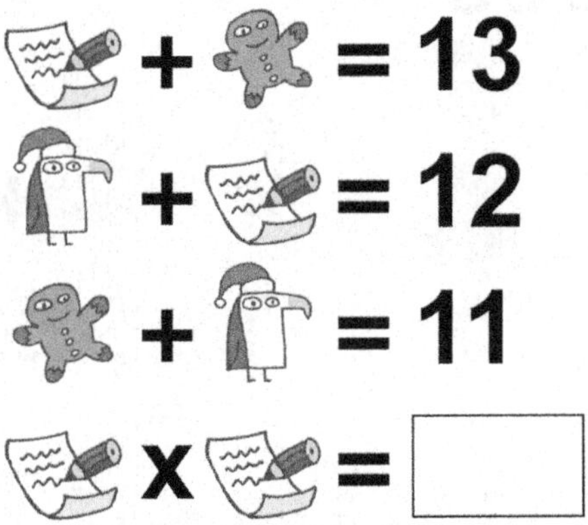

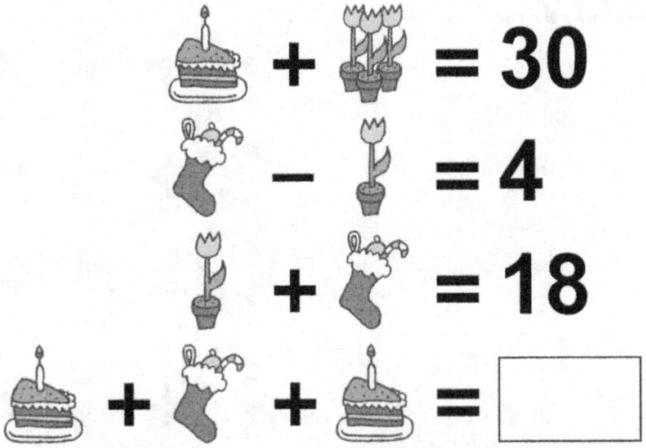

#211

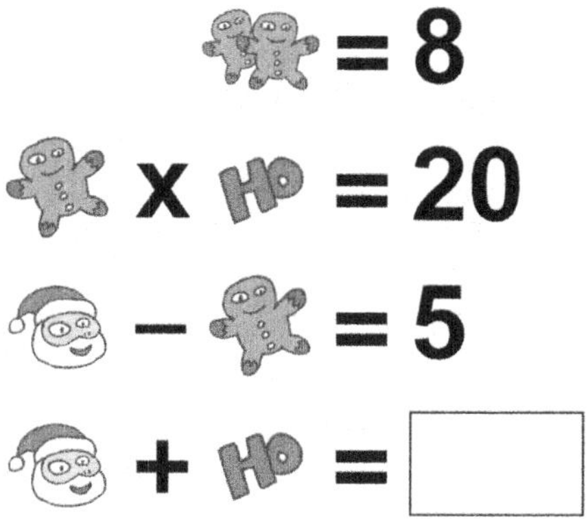

#212

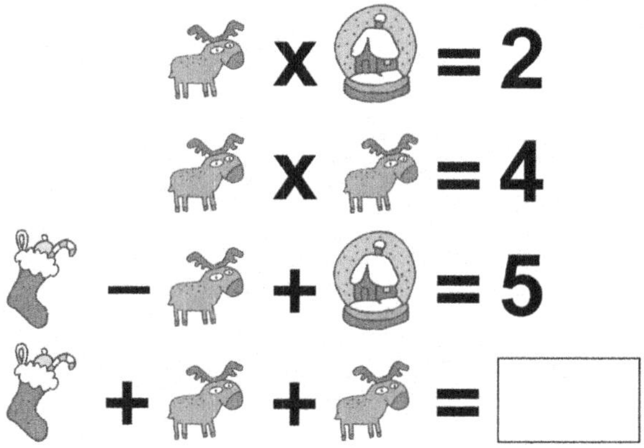

#213

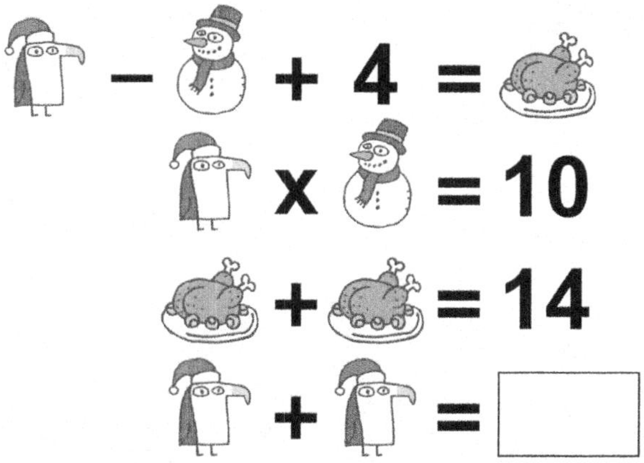

#214

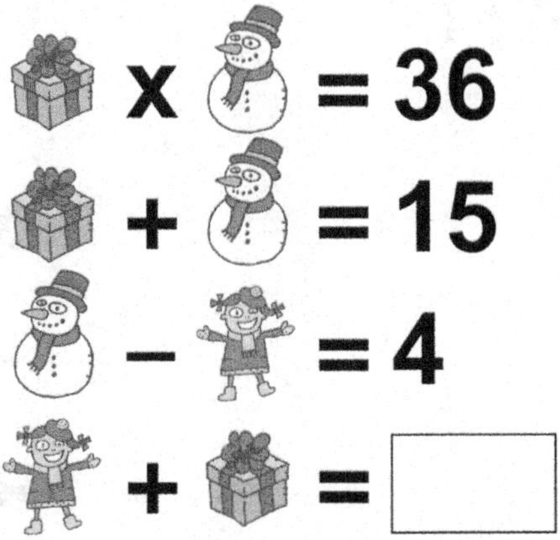

#215

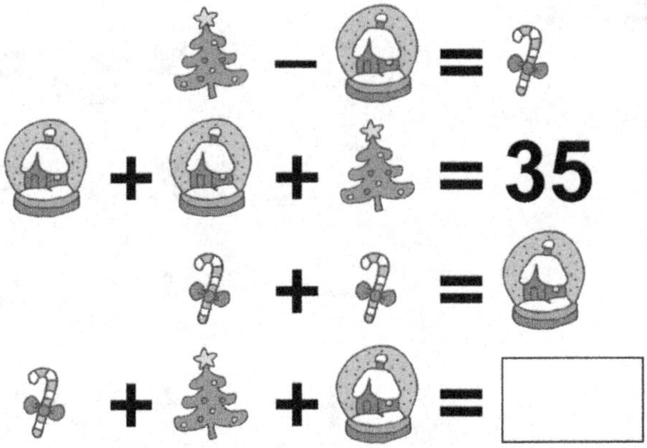

#216

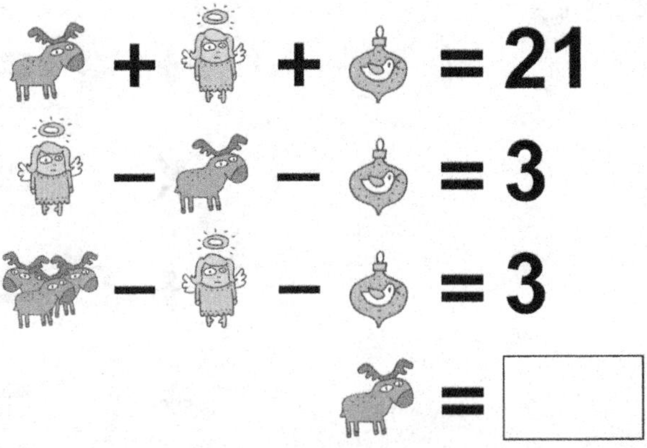

78

#217

🍪 + 🧦 = 8

🔔 + 🧦 = 17

🔔 + 🍪 = 15

🍪 + 🧦 + 🔔 = ☐

#218

🍗 − 🎄 = 🍬

🍬 + 🎄 = 20

🍗 − 🍬 = 15

🍗 − 🍬 − 🎄 = ☐

#219

🔮 x ⛄ = 20

🔮 + 🎅 = 4

🎅 + ⛄ = 23

⛄ + 🔮 = ☐

79

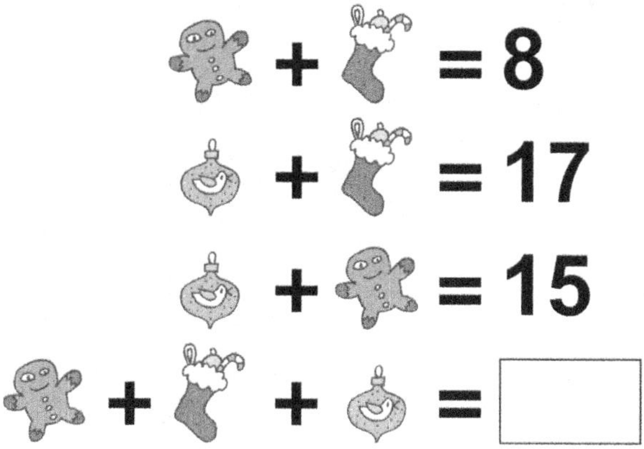

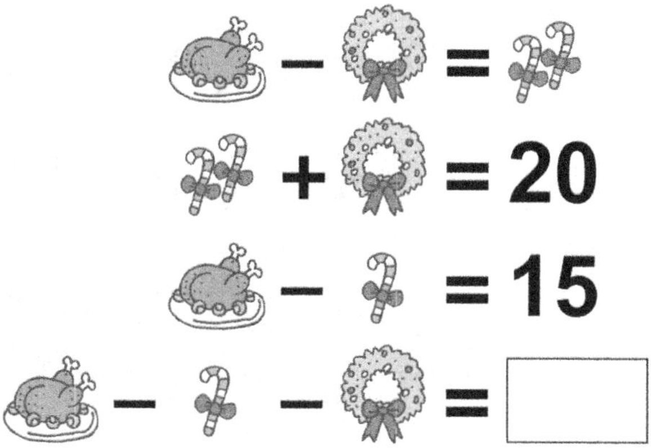

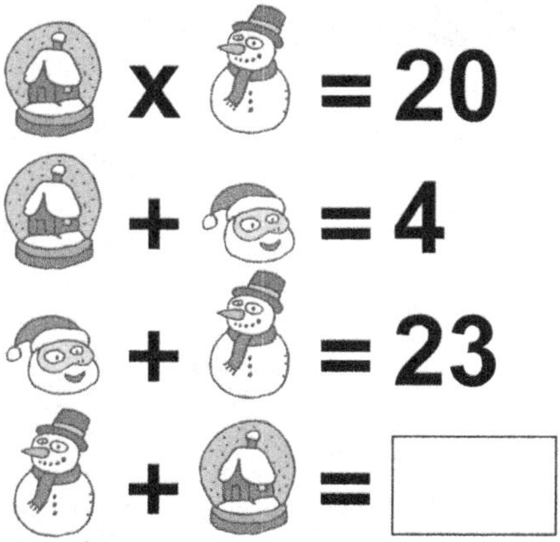

#220

HoHo − ❄ + 🐦 = 16

Ho + 🐦 + ❄ = 18

HoHo + 🐦🐦 = 24

❄ + ❄ = ☐

#221

🔥 + 🔥 + 📜 = 🎅

📜 + 📜 + 🔥 = 9

📜 − 🔥 = 🎅

🎅 + 🎅 + 🎅 = ☐

#222

👼 − 🎁 = 1

👻 + 🎁 = 11

👼 + 👼 − 🎁 = 15

🎁 × 👻 = ☐

80

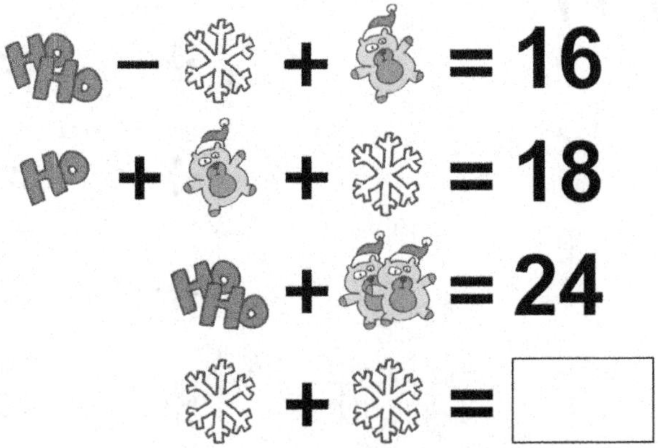

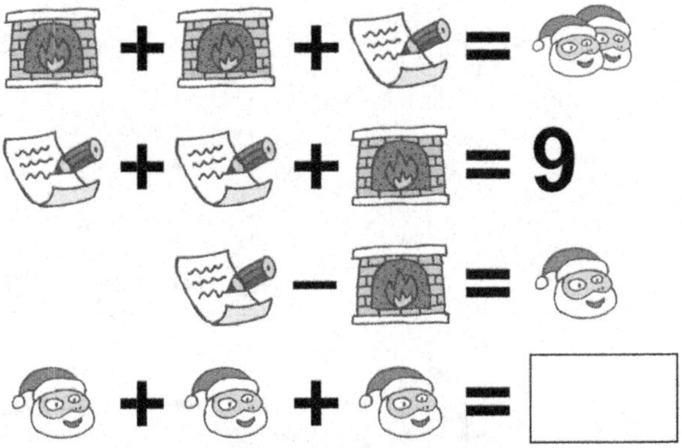

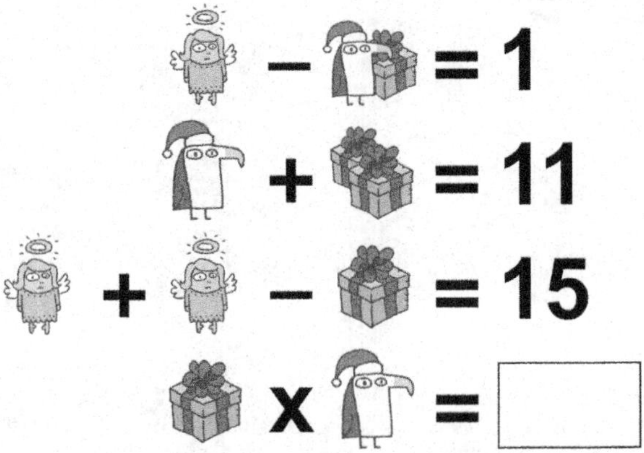

#223

 −  = 10

 −  = 6

 x = 12

 +  +  = ☐

#224

 −  = 3

 −  = 4

 − +  = 12

 + = ☐

#225

 x  =

 + = 40

 − = 1

 x = ☐

#226

🍬 + 🍬 = 🦌

🦌 − 🌷 − 🌷 = 2

🍬 + 🌷 = 9

🍬 − 🦌 = ☐

#227

❄ x 🎄 + 🎅 = 20

❄ − 🎄 = 3

3 x 🎅 = 24

🎅 + ❄ + ❄ = ☐

#228

🔔 + 🐦 = 🎁

🎁 x 🐦 = 18

👼 − 🎁 + 🔔 = 4

👼 + 👼 + 👼 = ☐

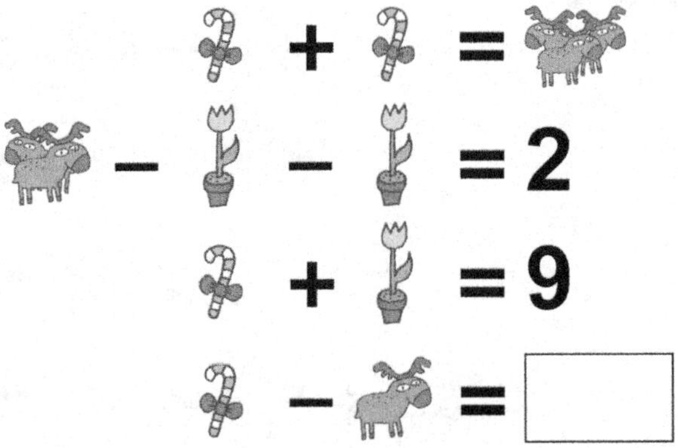

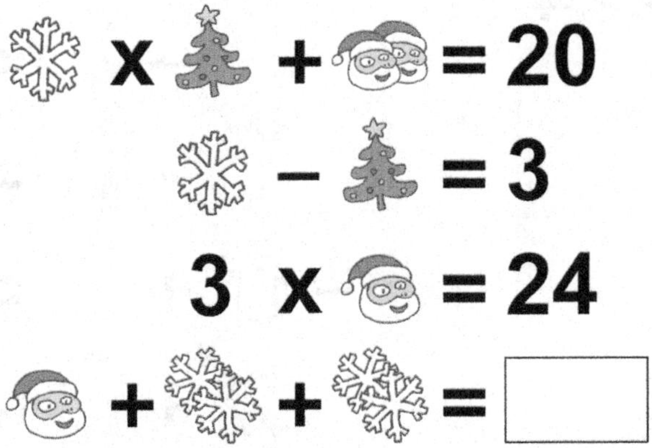

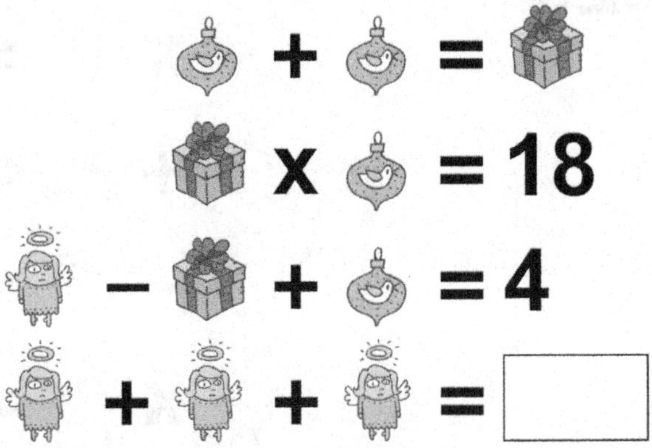

#229

 $-$  $= 1$

🦃 $\times$ HP $= 36$

HP $\times$ ⛄ $= 32$

🦃 $+$ HP $+$ ⛄ $=$ ____

#230

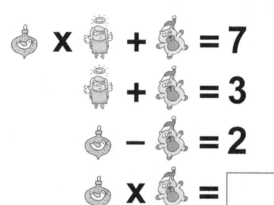

🔔 $\times$ 👼 $+$ 🎅 $= 7$

👼 $+$ 🎅 $= 3$

🔔 $-$ 🎅 $= 2$

🔔 $\times$ 🎅 $=$ ____

#231

🍪 $\times$ ❄ $+$ 🎅 $= 11$

🎅 $+$ ❄ $= 3$

🧝 $\times$ 🍪 $= 18$

🍪 $+$ 🧝 $=$ ____

#232

🍪 − 🐦 = 2

🍪 + 📃 − 🐦 = 8

🍪 + 🐦 = 16

📃 × 🐦 = ☐

#233

🦌 + 🍗 = 11

🐷🐷 − 🍗 = 7

🔥 + 6 − 🦌 = 5

🔥 + 🔥 + 🔥 = ☐

#234

🧝 + ⛄ = 🦌

🦌 + 🧝 − ⛄ = 16

🦌 × ⛄ = 10

🧝 − 🦌 = ☐

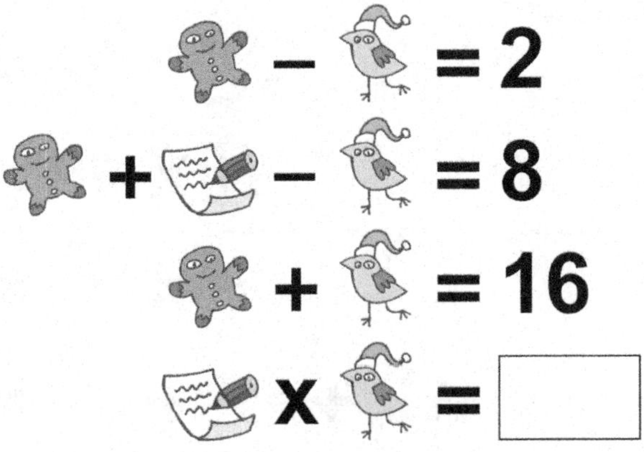

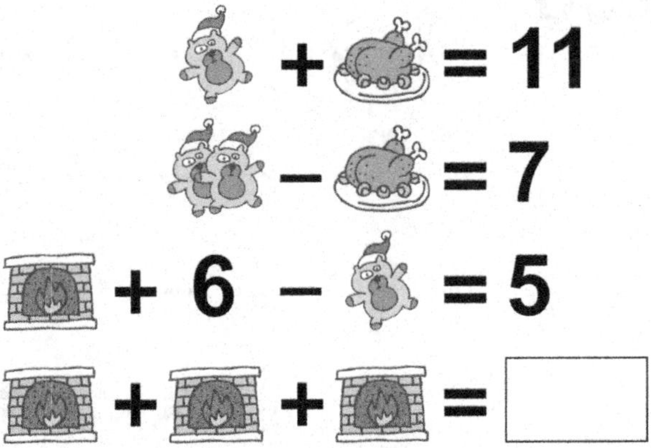

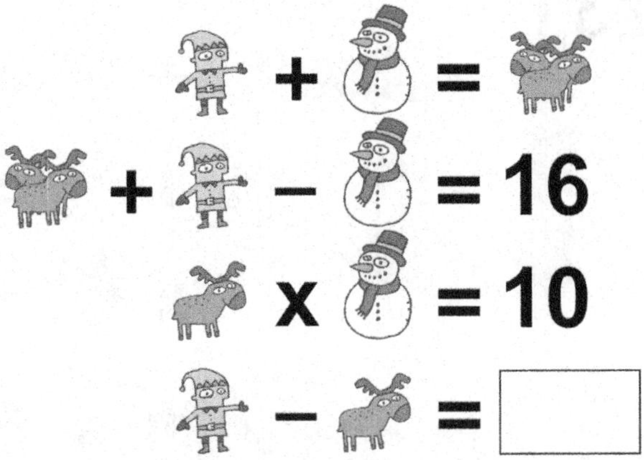

#235

 x x = 9

 − = 2

 x = 12

 + + = ☐

#236

 − = 11

 x = 20

 +  = 14

 x = ☐

#237

 x  − = 13

 +  = 8

 x  = 25

 x = ☐

#238

🌷 + 🌷🌷🌷 + 🍗 = 15

❄️ − 🍗 = 🌷

❄️ x ❄️ = 81

🍗 − 🌷 = ☐

#239

👧 x 🧝🎄 + 🎂 = 500

🎄 + 🧝 = 100

🎂 − 👧 = 197

🧝🎄 − 👧 = ☐

#240

❄️ + 🧦 = 15

❄️ x 🧦 = 54

📝 + 📝 − 🧦 = 7

📝 − ❄️ = ☐

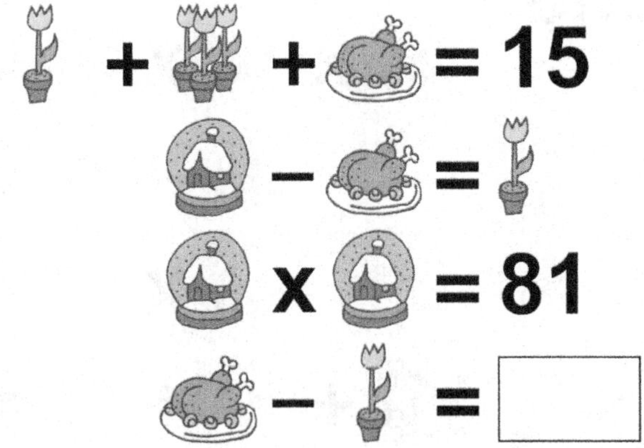

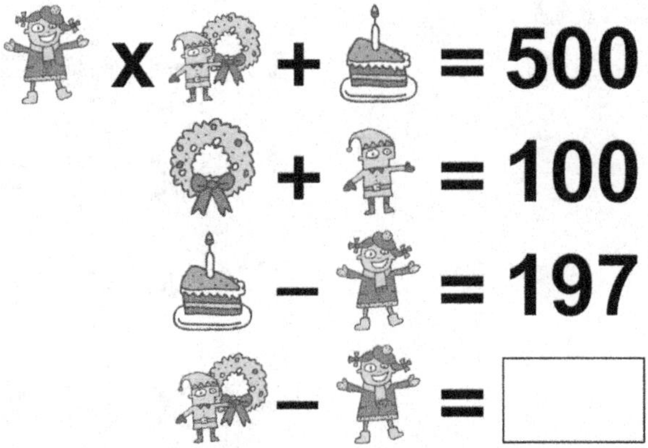

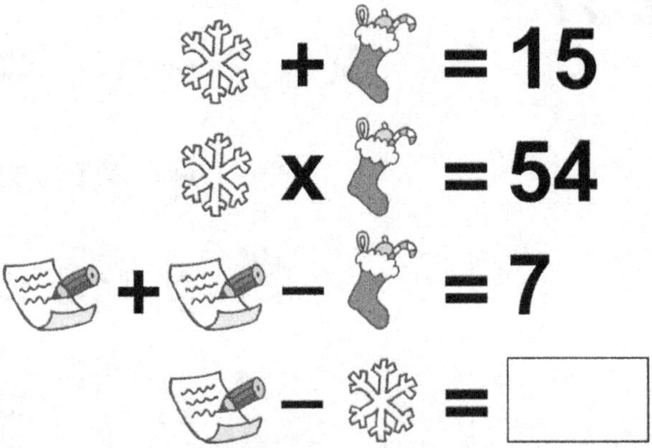

#241

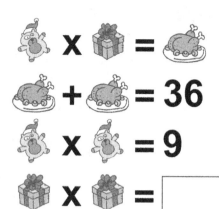

$$\text{(santa)} \times \text{(gift)} = \text{(turkey)}$$
$$\text{(turkey)} + \text{(turkey)} = 36$$
$$\text{(santa)} \times \text{(santa)} = 9$$
$$\text{(gift)} \times \text{(gift)} = \boxed{}$$

#242

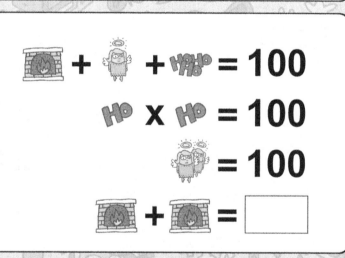

$$\text{(fireplace)} + \text{(angel)} + \text{(hohoho)} = 100$$
$$\text{(ho)} \times \text{(ho)} = 100$$
$$\text{(angels)} = 100$$
$$\text{(fireplace)} + \text{(fireplace)} = \boxed{}$$

#243

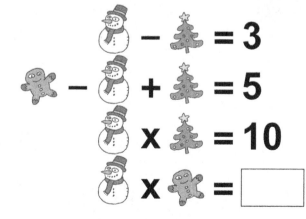

$$\text{(snowman)} - \text{(tree)} = 3$$
$$\text{(gingerbread)} - \text{(snowman)} + \text{(tree)} = 5$$
$$\text{(snowman)} \times \text{(tree)} = 10$$
$$\text{(snowman)} \times \text{(gingerbread)} = \boxed{}$$

#244

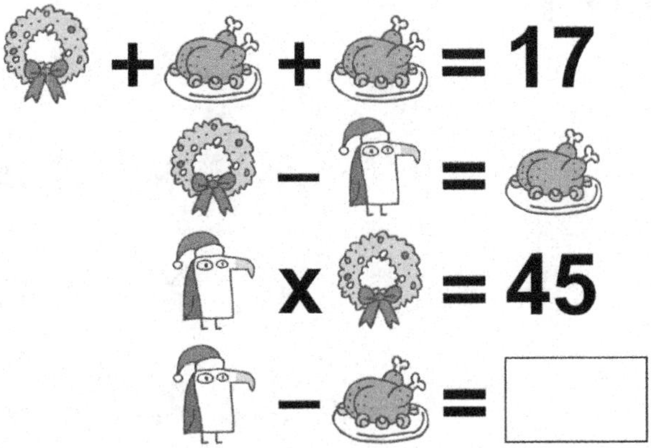

#245

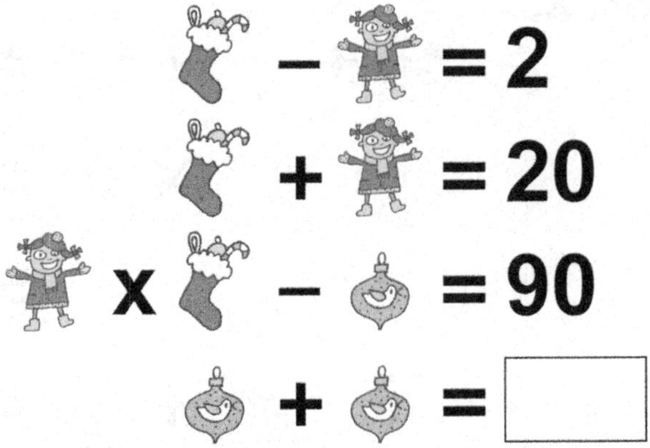

#246

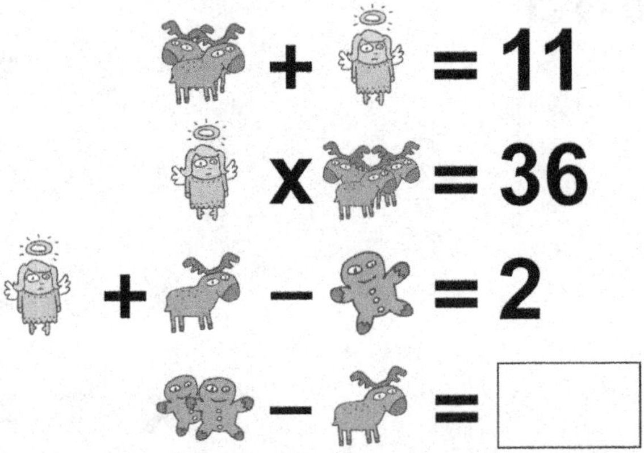

88

#247

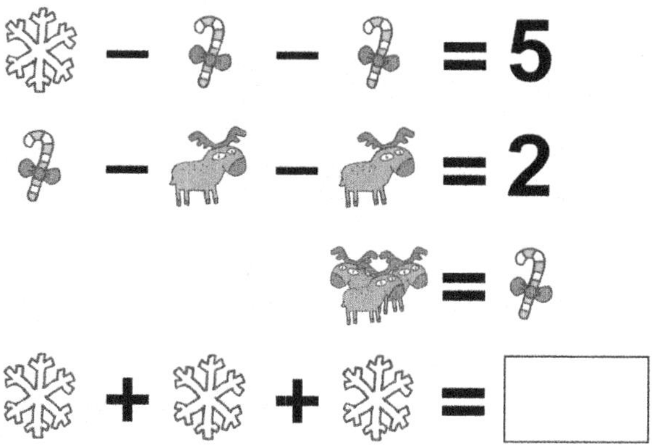

#248

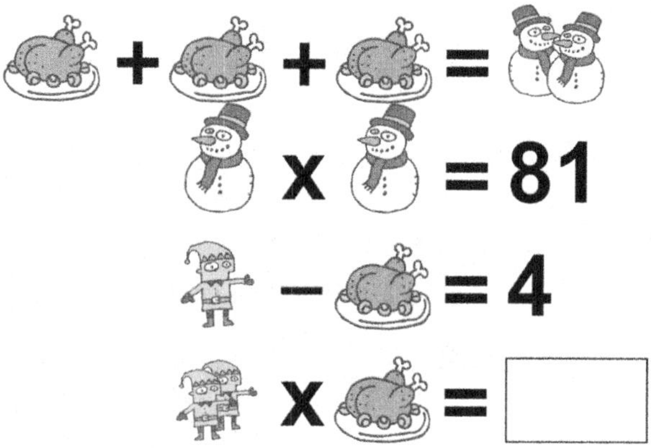

#249

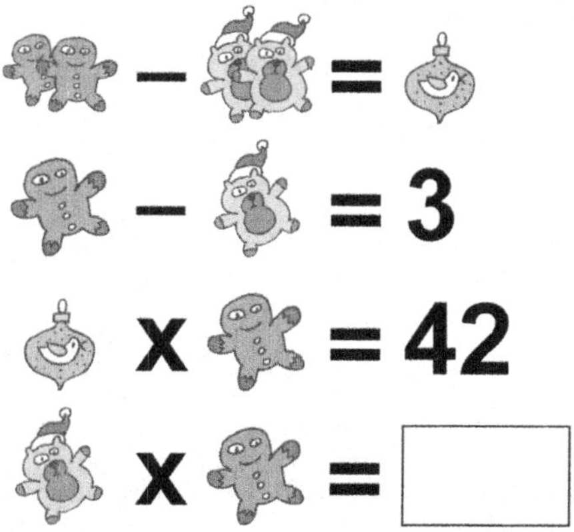

#250

cake − tulip = 2

tulip × cake − HO = 10

HO − cake = 0

HO + tulip = ☐

#251

reindeer + santa + fireplace = 58

santa − reindeer = fireplace

reindeer + fireplace + reindeer = 52

santa + santa = ☐

#252

elf × snowflake = 100

snowflake − elf = 48

snowflake + girl + girl = 74

girl + elf + elf = ☐

90

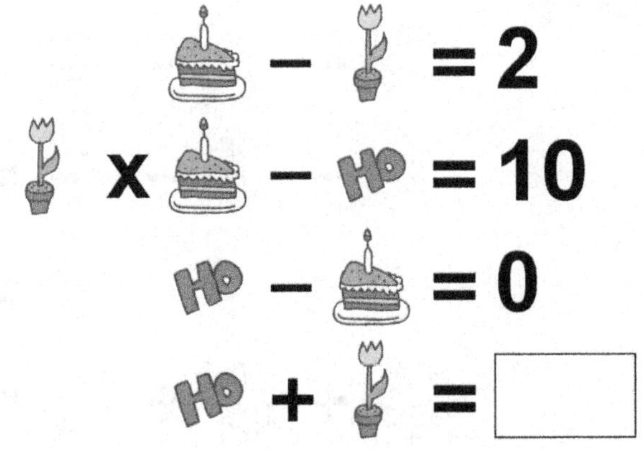

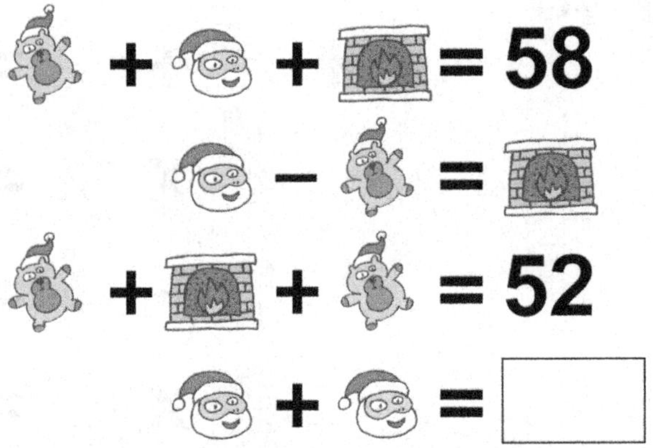

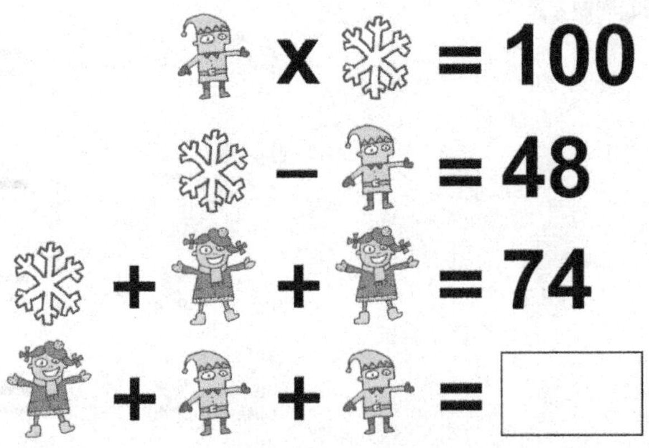

#253

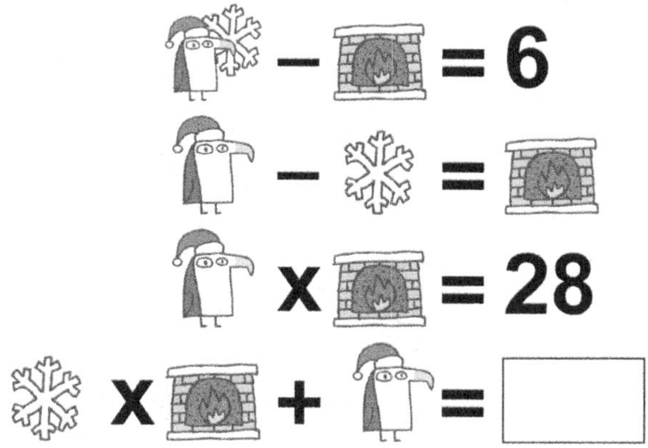

#254

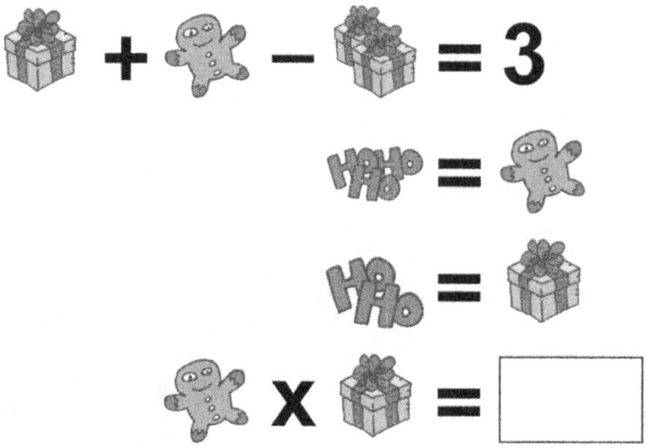

#255

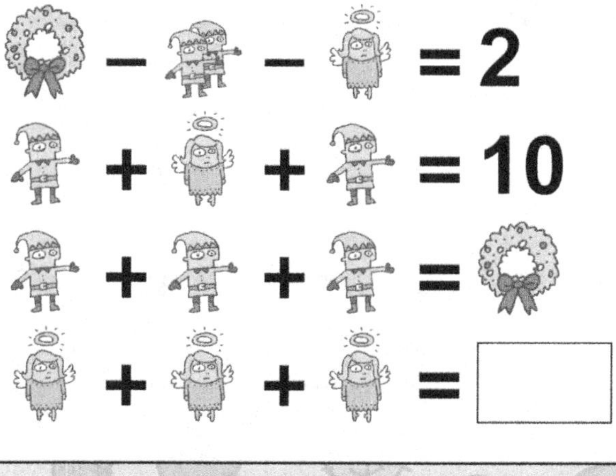

91

#256

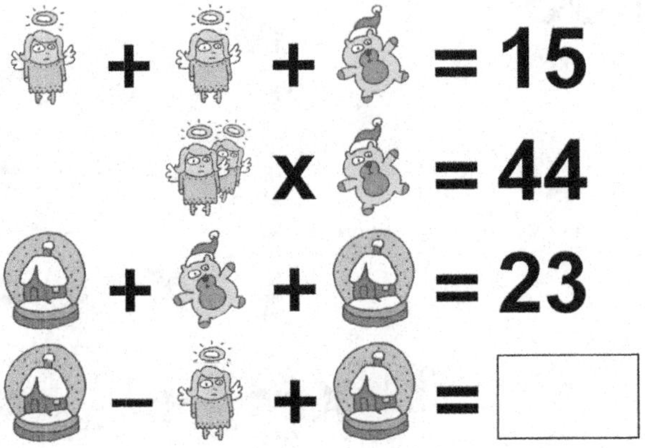

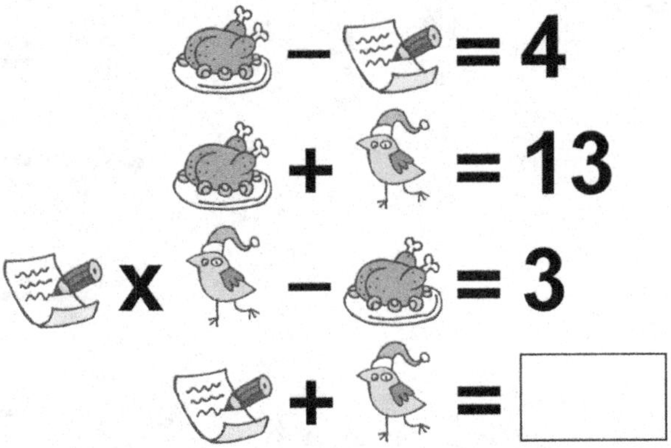

🌨️ + 🐦 + 🌨️ = 23

🌨️ − 😇 + 🌨️ = ▢

#257

🍗 − 📜 = 4

🍗 + 🐦 = 13

📜 x 🐦 − 🍗 = 3

📜 + 🐦 = ▢

#258

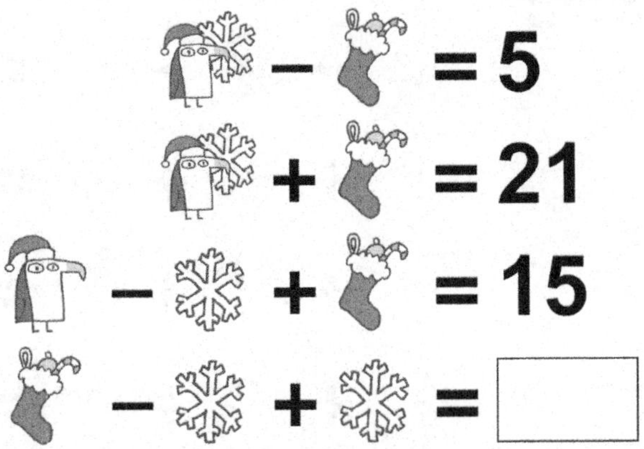

🧙 − ❄️ + 🧦 = 15

🧦 − ❄️ + ❄️ = ▢

#259

🎁 + 🐷 + 🔥 = 15

🎁 + 🔥 = 🐷🐷

🐷 − 🎁 = 1

🐷 x 🎁 x 🔥 = ⬜

#260

🎅 − 🔔 = 4

🔮 − 🎅 − 🔔 = 5

🔮 − 🔔 − 🔔 = 🎅

🔮 + 🎅 + 🔔 = ⬜

#261

🧦 x 🧦 = 32

🦌 x 🧦 = 24

🦌 + 🍗 = 15

🍗 + 🍗 + 🍗 = ⬜

#262

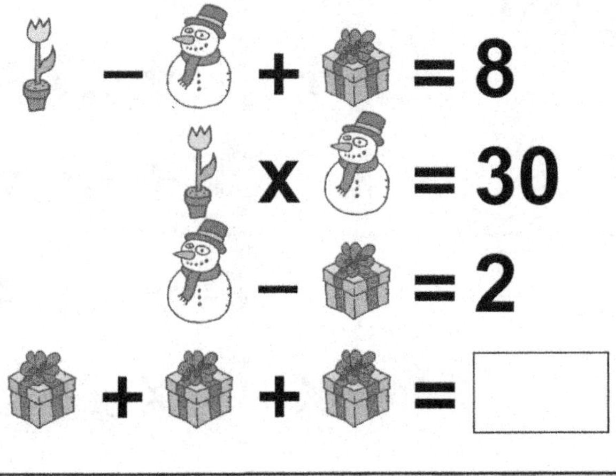

🌷 − ☃ + 🎁 = 8

🌷 x ☃ = 30

☃ − 🎁 = 2

🎁 + 🎁 + 🎁 = ☐

#263

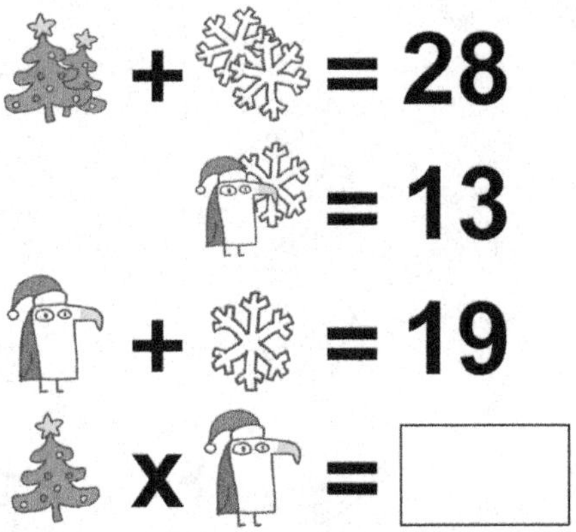

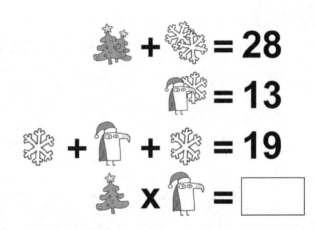

🎄 + ❄ = 28

🎅 = 13

❄ + 🎅 + ❄ = 19

🎄 x 🎅 = ☐

#264

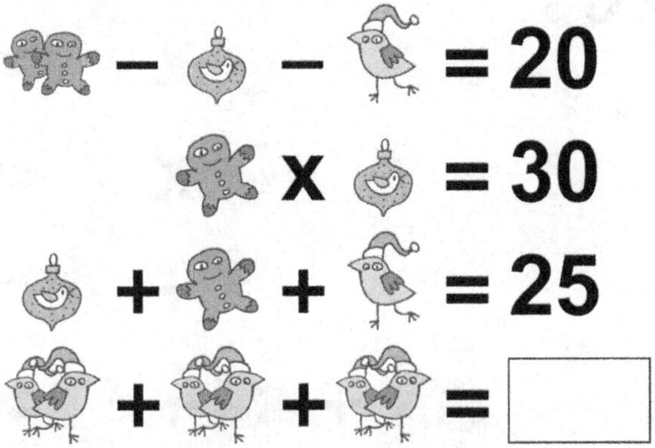

🍪 − 🔔 − 🐦 = 20

🍪 x 🔔 = 30

🔔 + 🍪 + 🐦 = 25

🐦 + 🐦 + 🐦 = ☐

#265

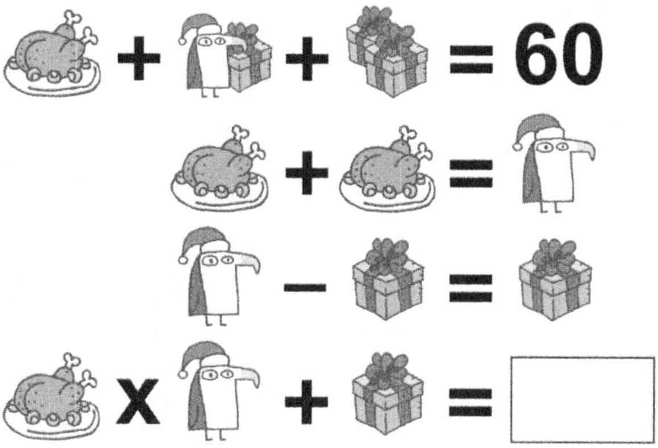

#266

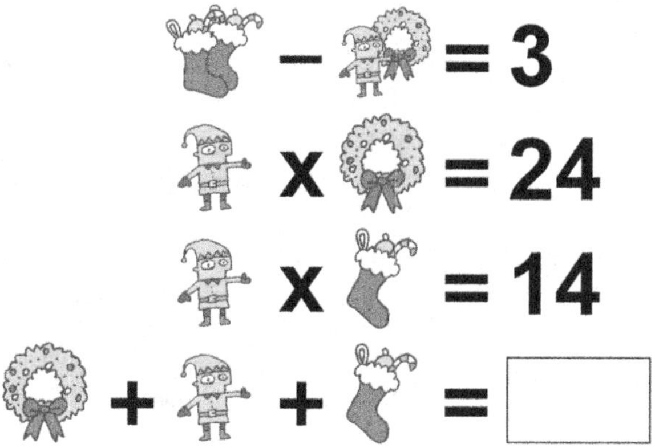

#267

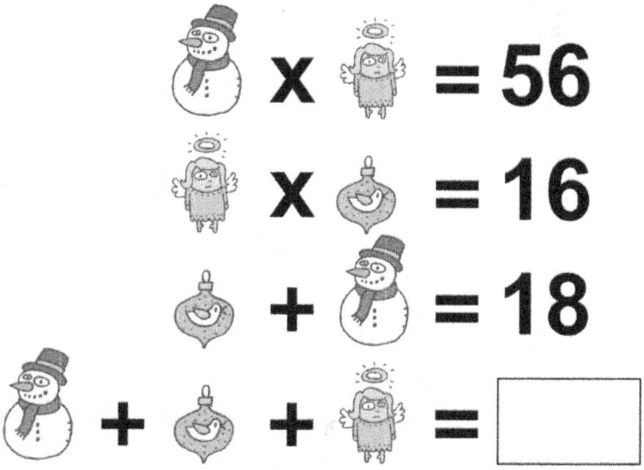

#268

🎂 + HO = 12

HOHO − 🎄 − 🎂 = 4

HO + 🎄 = 9

🎂 − 🎄 = ☐

#269

☃ x 🎄🎁 + 🎁 = 29

🎁 + 🧝 = ☃

🌷 x 🎁 = 28

🧝 + 🌷 + 🧝 = ☐

#270

🍬 + 🔔 = 13

🧒 + 🔔 = 🍬

🔔 − 🧒 + 7 = 12

🧒 + 🍬 + 🧒 = ☐

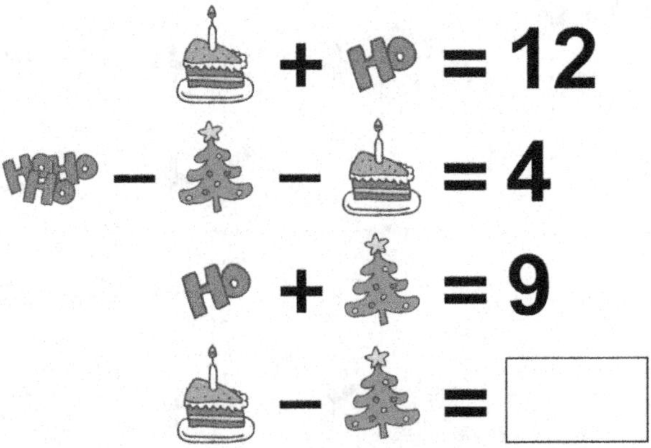

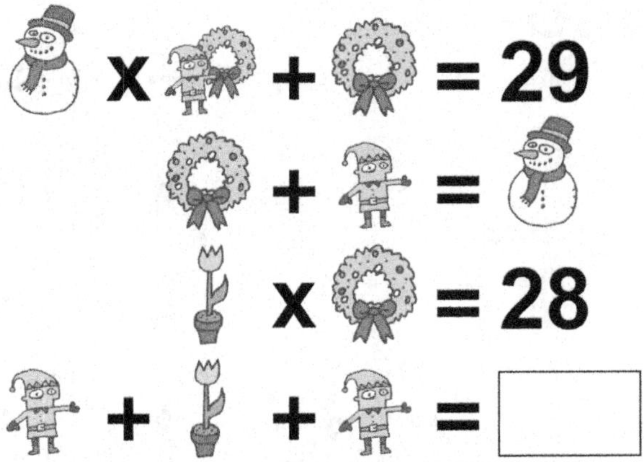

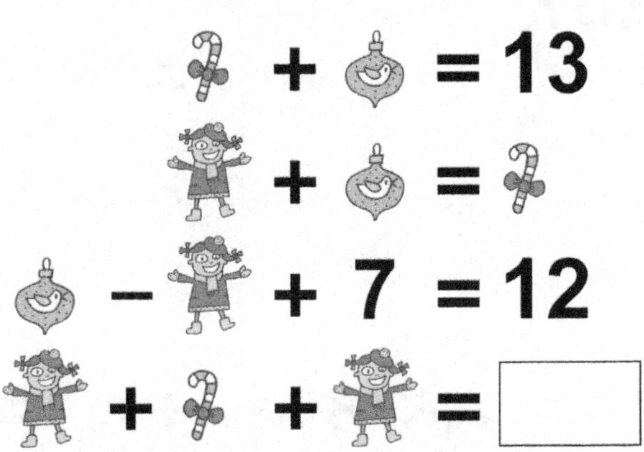

#271

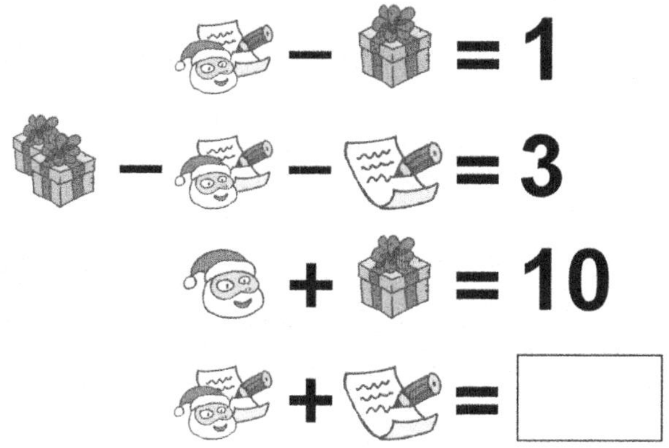

$$\text{🎅} - \text{🎁} = 1$$
$$\text{🎁} - \text{🎅} - \text{📜} = 3$$
$$\text{🎅} + \text{🎁} = 10$$
$$\text{🎅} + \text{📜} = \boxed{}$$

#272

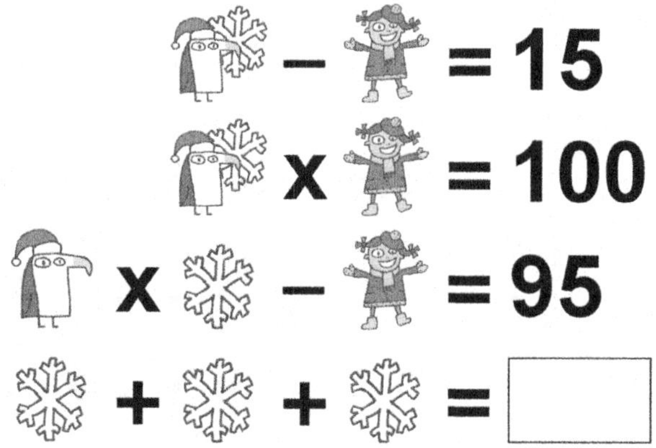

$$\text{🧝} - \text{🧒} = 15$$
$$\text{🧝} \times \text{🧒} = 100$$
$$\text{🧝} \times \text{❄} - \text{🧒} = 95$$
$$\text{❄} + \text{❄} + \text{❄} = \boxed{}$$

#273

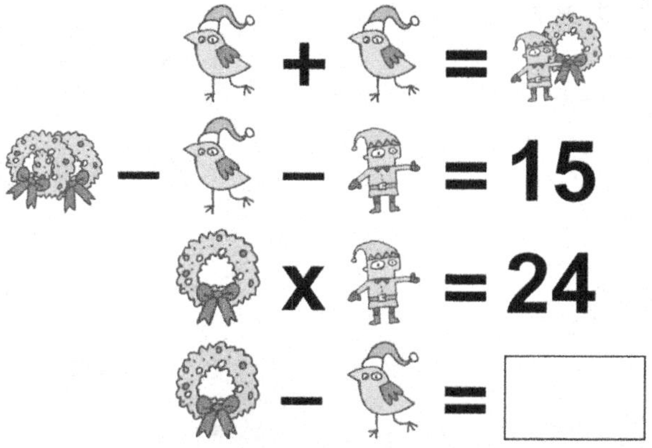

$$\text{🐦} + \text{🐦} = \text{🧝🎄}$$
$$\text{🎄} - \text{🐦} - \text{🧝} = 15$$
$$\text{🎄} \times \text{🧝} = 24$$
$$\text{🎄} - \text{🐦} = \boxed{}$$

#274

turkey $+$ angel $-$ stocking $= 30$

turkey $+$ stocking $-$ angel $= 6$

angel $\times$ stocking $= 45$

turkey $-$ angel $=$ ☐

#275

cake $+$ HO $= 8$

cake $\times$ HO $=$ moose-pair

moose $+$ cake $= 8$

moose $+$ HO $=$ ☐

#276

elf $+$ fireplace $= 75$

$4 \times$ elf $=$ fireplace

elf $-$ santa $= 4$

fireplace $-$ santa $=$ ☐

98

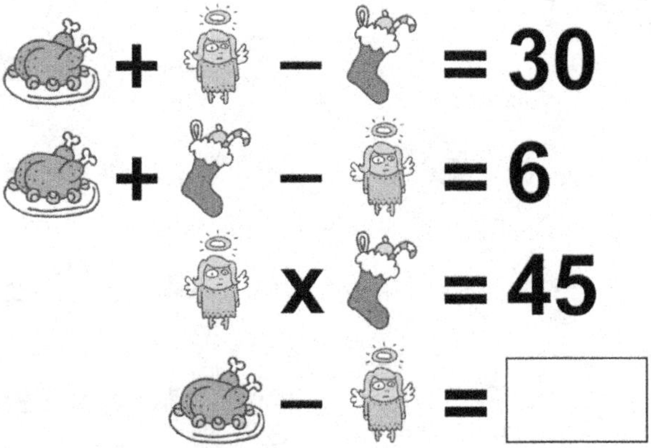

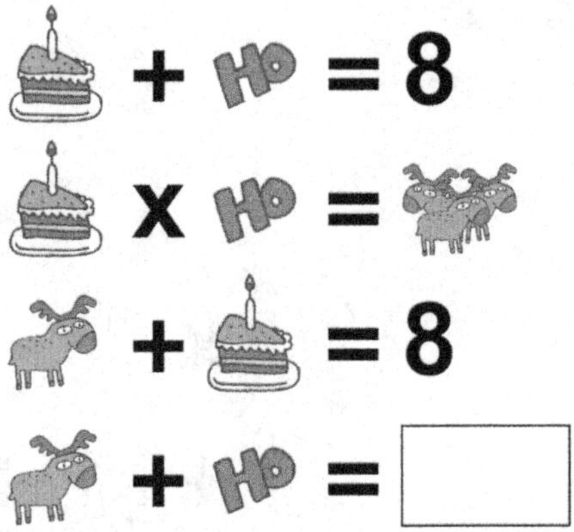

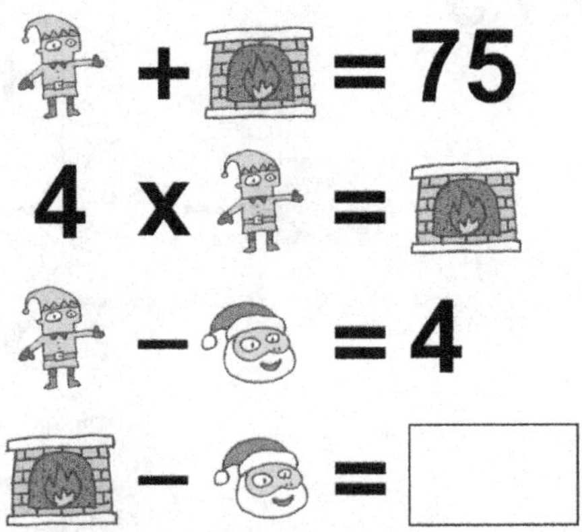

#277

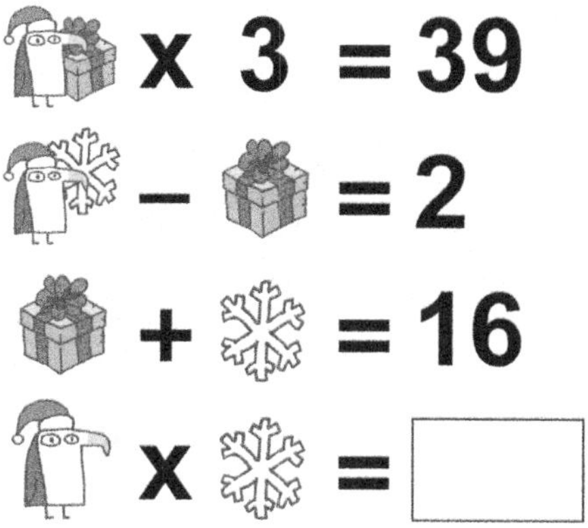

#278

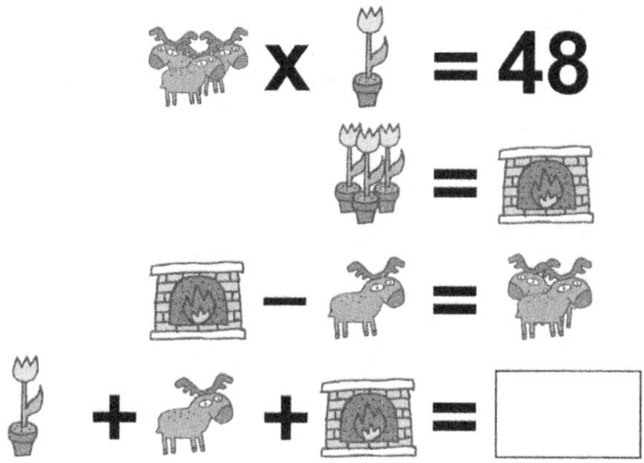

#279

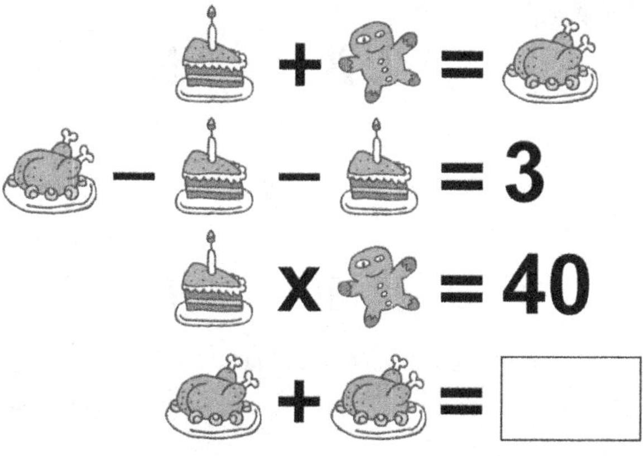

99

#280

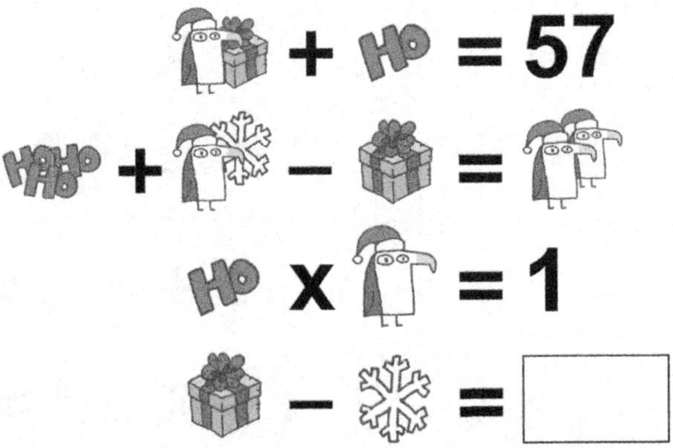

#281

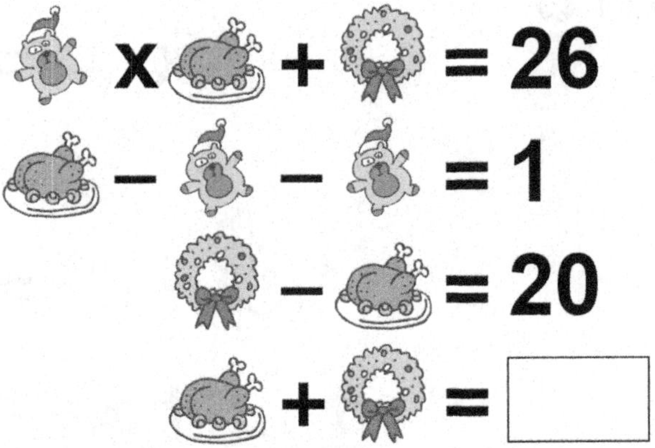

#282

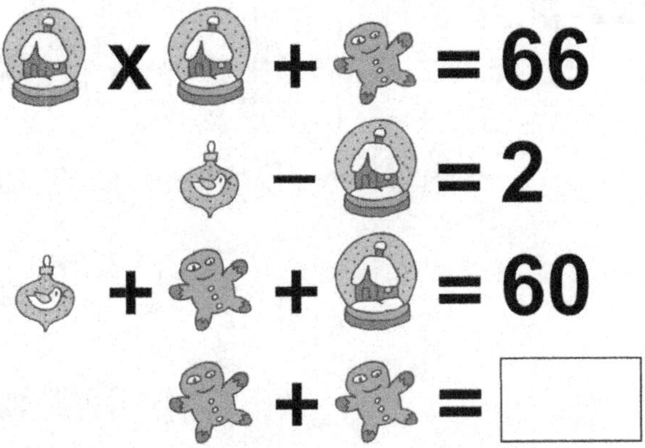

#283

 + = 17

🎅 − 🎄 = 1

🎅 x 🧝 = 49

📜 + 🎄 = ▭

#284

🔔 x 👼 = 🧦

🧦 + 👼 = 32

🧦 − 🔔 = 21

👼 + 🔔 + 🧦 = ▭

#285

🐶 + ⛄ = HOHO

⛄⛄ − HO − 🐶 = 10

HOHOHO = 18

🐶 x ⛄ = ▭

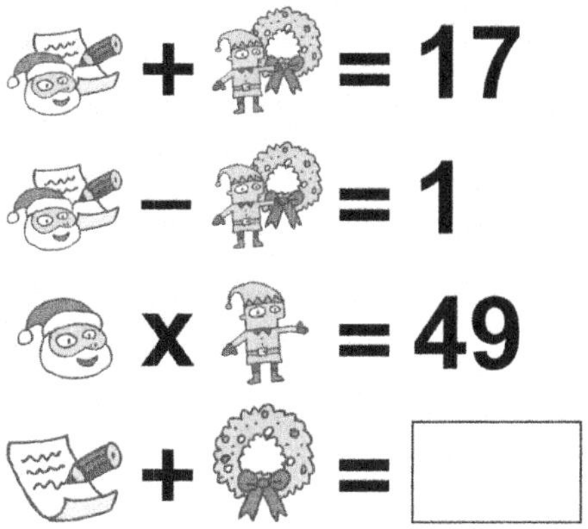

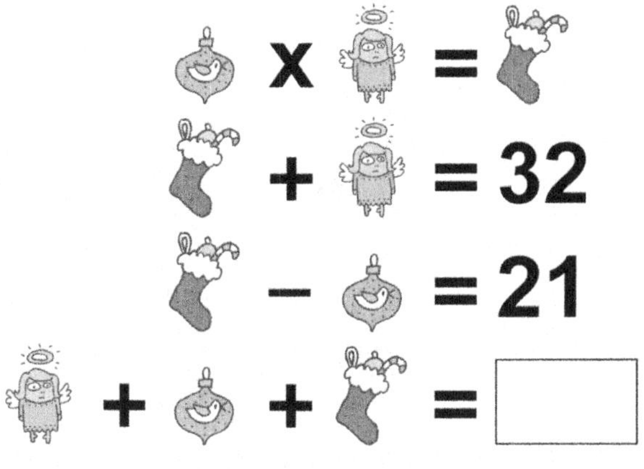

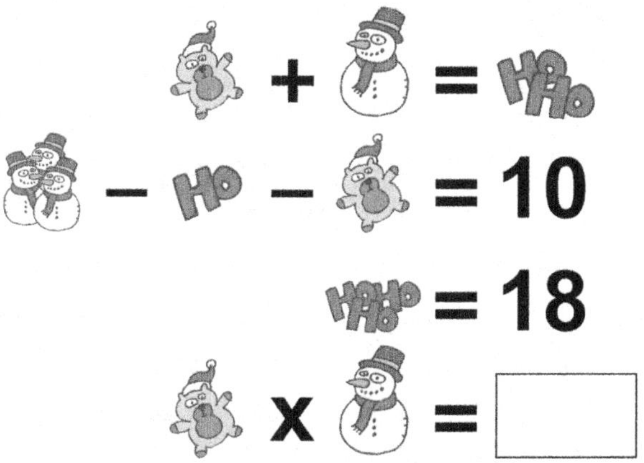

#286

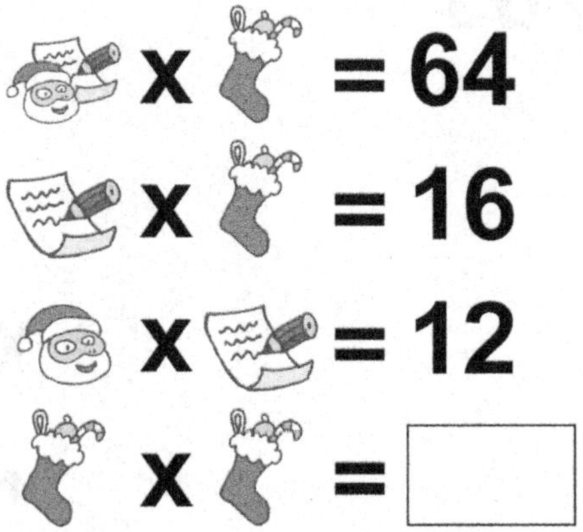

Santa × stocking = 64

list × stocking = 16

Santa × list = 12

stocking × stocking = ☐

#287

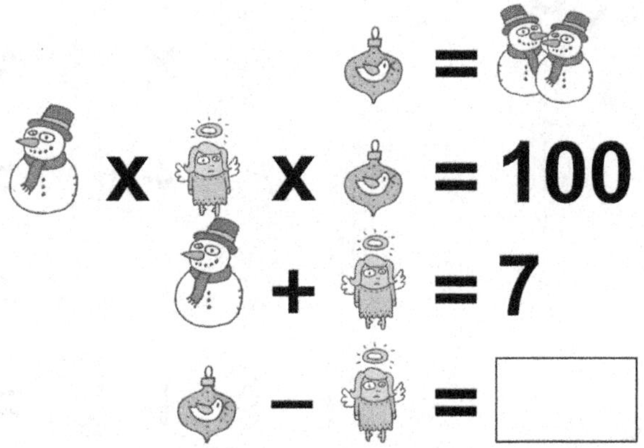

ornament = two snowmen

snowman × angel × ornament = 100

snowman + angel = 7

ornament − angel = ☐

#288

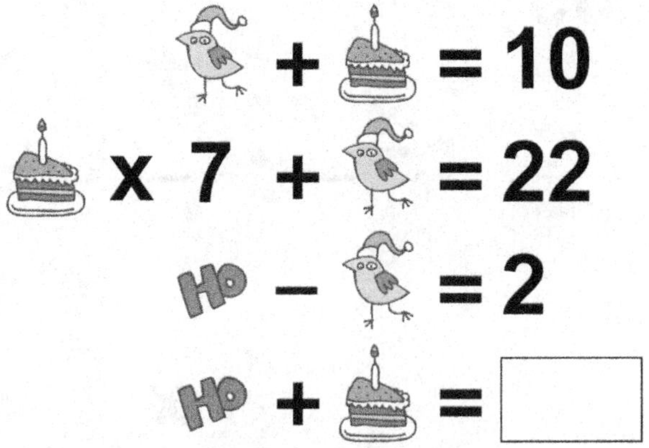

bird + cake = 10

cake × 7 + bird = 22

HO − bird = 2

HO + cake = ☐

#289

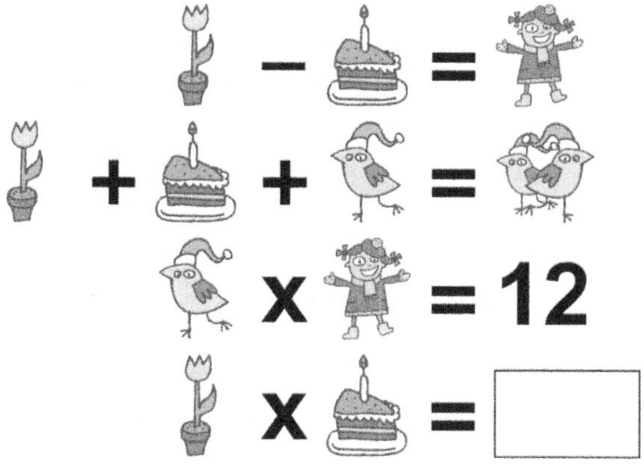

#290

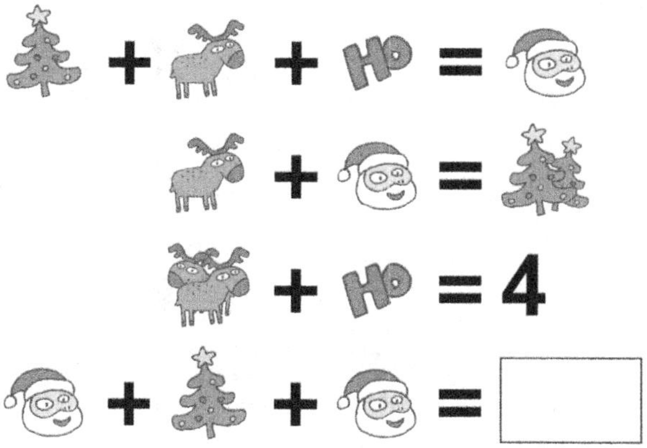

#291

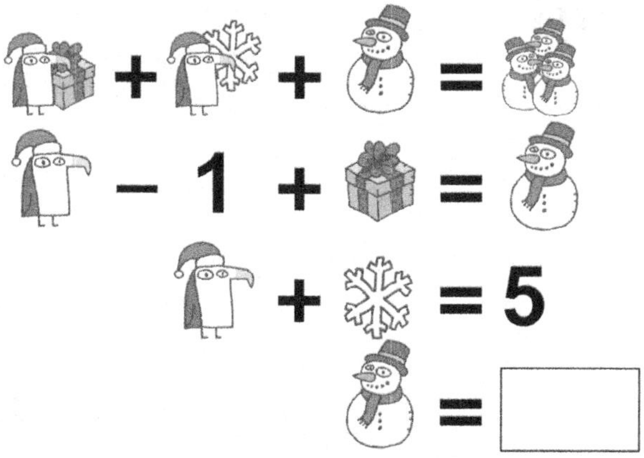

#292

🍬 + 🔥 + 👧 = 🍬🍬

🍬 × 🎀 = 8

🍬 − 🔥 − 🎀 = 2

👧 × 🎀 = ☐

#293

🎅 × 🎅 = 50

📜 × 🎅🎅 = 50

⛄ + 📜 = 12

⛄⛄ − 🎅 − 🎅 = ☐

#294

❄ × 🧝 = 👼👼

🧝 − ❄ = 7

👼 − 🧝 = 5

👼 + ❄ = ☐

104

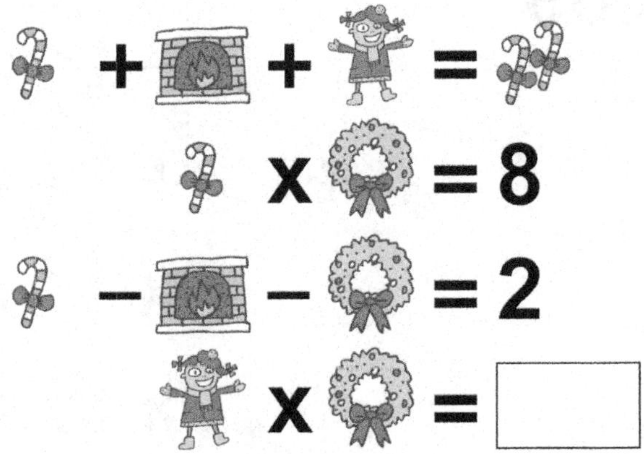

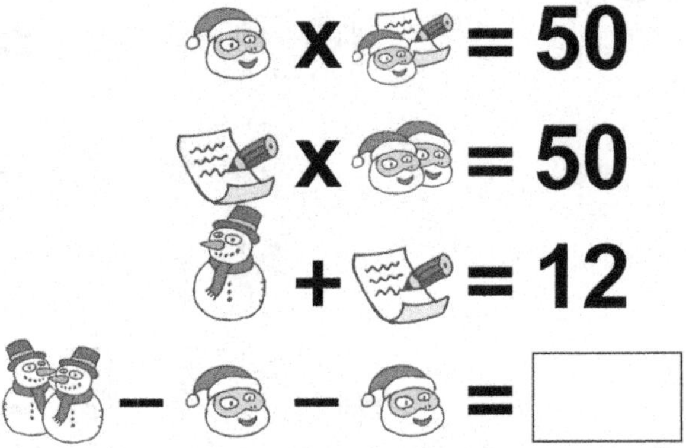

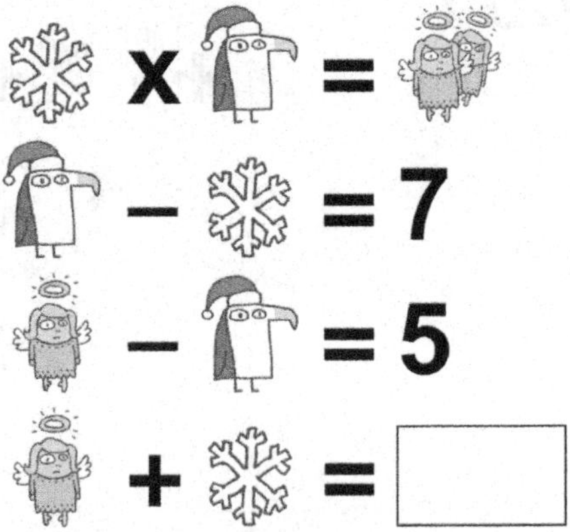

#295

🧦 − 👼👼 = 🦌

🦌 + 👼 + 🧦 = 17

🧦 − 👼 = 3

🦌🦌 = ⬜

#296

🧝 − 50 = ❄️

🎄 + 🧝 = 100

❄️ x 🎄 = 200

❄️ + ❄️ + ❄️ = ⬜

#297

🎅 + HO + 🐦 = 39

🍗 − 🐦 + 11 = 🎅

🐦 = 7

🍗 + HO = ⬜

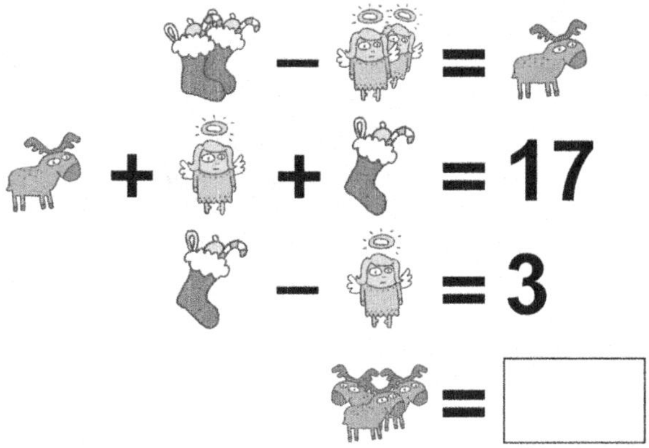

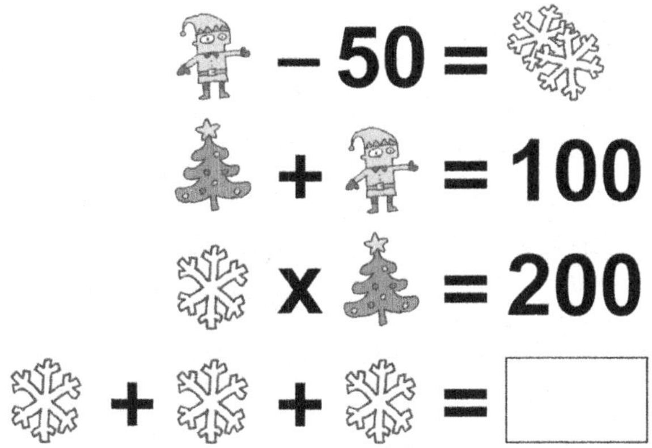

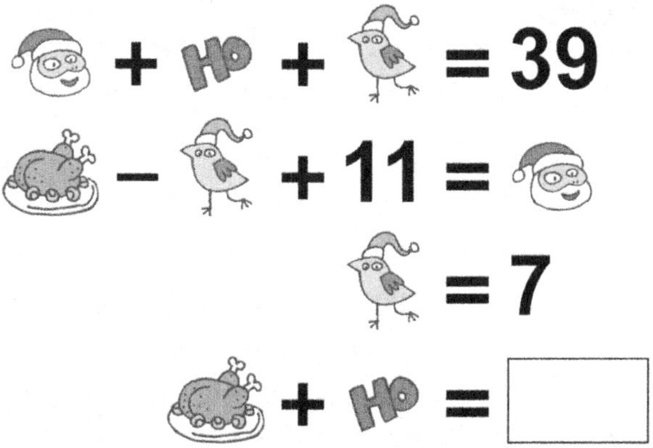

#298

 = 53

 = 47

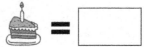 = 50

 =

#299

 −  = 3

 − 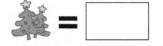 = 2

 =

#300

 = 42

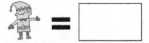 =

FINISH

Good job on making it to the end! Now it's time to see if you got everything correct. Flip to the next page and start checking your answers.

If you enjoyed this book, you can find similar titles on my Amazon Author Page. Just scan the code!

https://www.amazon.com/J-J-Wiggins/e/B01AWMH7BU

Don't forget to leave a review on Amazon.

Thank you!

ANSWERS

#1: 8   5   2   1
#2: 10   3   2   5
#3: 6   5   3   4
#4: 6   1   1   4
#5: 2   4   2   10
#6: 1   1   5   4
#7: 12   4   4   4
#8: 13   10   5   1
#9: 12   2   9   4
#10: 4   1   5   5
#11: 5   1   2   4
#12: 8   5   *   *
#13: 6   4   2   3
#14: 1   5   2   2
#15: 5   5   3   8
#16: 1   1   4   2
#17: 12   4   3   5
#18: 1   2   3   4
#19: 15   2   5   3
#20: 6   3   3   5
#21: 34   17   4   3
#22: 3   1   5   10
#23: 3   3   2   5
#24: 14   5   4   2
#25: 2   3   1   5
#26: 24   2   3   8

#27: 3   2   5   7
#28: 45   4   7   15
#29: 1   5   2   3
#30: 6   2   3   1
#31: 1   5   1   3
#32: 24   8   3   6
#33: 3   9   3   1
#34: 9   4   3   2
#35: 11   5   12   3
#36: 2   4   5   3
#37: 11   3   5   7
#38: 3   8   6   3
#39: 6   4   3   5
#40: 15   9   4   2
#41: 1   4   3   5
#42: 1   9   5   4
#43: 7   3   6   5
#44: 1   10   3   7
#45: 18   1   9   5
#46: 10   1   3   4
#47: 6   3   5   3
#48: 3   11   7   6
#49: 5   3   2   4
#50: 6   8   4   7
#51: 90   5   30   5
#52: 7   3   6   1

#53: 3   4   5   7
#54: 4   2   5   3
#55: 100   50   10   20
#56: 13   5   9   3
#57: 2   3   1   6
#58: 9   4   2   1
#59: 12   5   3   4
#60: 5   1   7   3
#61: 17   11   10   4
#62: 2   2   3   1
#63: 5   5   4   1
#64: 16   9   7   1
#65: 3   5   4
#66: 4   6   7   10
#67: 27   4   3   9
#68: 4   3   2   4
#69: 15   3   4   8
#70: 19   4   5   10
#71: 6   3   2   5
#72: 19   7   4   4
#73: 21   7   4   7
#74: 21   7   5   7
#75: 25   20   7   2
#76: 29   9   6   7
#77: 10   8   1   2
#78: 12   4   10   1
#79: 94   98   3   1
#80: 38   15   12   50
#81: 4   4   2   9
#82: 2   2   10   8
#83: 4   8   3   4

#84: 39   3   7   12
#85: 225   50   49   25
#86: 18   10   7   4
#87: 17   9   5   6
#88: 28   8   7   10
#89: 26   6   3   2
#90: 28   6   10   5
#91: 31   7   13   25
#92: 1   4   5   2
#93: 50   13   10   7
#94: 8   6   3   5
#95: 5   4   2   7
#96: 5   4   3   7
#97: 10   4   1   5
#98: 22   8   5   9
#99: 51   50   55   4
#100: 3   3   3   1
#101: 24   10   5   8
#102: 3   3   1   6
#103: 18   6   12   3
#104: 3   3   6   9
#105: 8   8   10   2
#106: 8   8   4   16
#107: 5   4   1   2
#108: 90   19   11   30
#109: 59   7   7   52
#110: 1   4   8   7
#111: 2   6   4   2
#112: 3   5   1   4
#113: 5   8   3   13
#114: 6   2   1   5

#115: 3   4   6   9
#116: 24   10   5   7
#117: 8   5   1   3
#118: 6   5   2   9
#119: 10   4   4   2
#120: 15   17   6   2
#121: 22   6   8   4
#122: 32   4   6   16
#123: 4   10   4   6
#124: 12   10   4   6
#125: 17   5   1   6
#126: 48   8   5   24
#127: 3   2   1   1
#128: 9   6   9   3
#129: 9   3   4   4
#130: 24   19   5   4
#131: 57   19   4   9
#132: 4   8   9   5
#133: 18   2   5   6
#134: 13   4   2   7
#135: 10   8   9   2
#136: 4   2   4   8
#137: 8   9   4   5
#138: 9   7   2   2
#139: 50   29   25   8
#140: 9   7   3   5
#141: 30   7   5   10
#142: 6   3   3   4
#143: 15   3   6   9
#144: 60   60   30   90
#145: 9   7   5   1

#146: 8   7   3   6
#147: 6   5   3   4
#148: 36   5   12   7
#149: 2   2   4   6
#150: 30   15   10   5
#151: 13   5   6   8
#152: 3   7   1   6
#153: 5   5   4   14
#154: 11   6   2   3
#155: 14   14   14   7
#156: 30   30   90   18
#157: 4   7   3   5
#158: 12   7   1   6
#159: 1   6   2   9
#160: 17   10   3   11
#161: 14   1   7   8
#162: 6   6   36   18
#163: 2   4   2   4
#164: 1   6   5   7
#165: 42   8   8   7
#166: 8   2   3   10
#167: 25   10   5   15
#168: 66   15   6   17
#169: 6   6   8   3
#170: 36   6   3   9
#171: 20   10   6   6
#172: 22   5   1   6
#173: 4   9   2   18
#174: 3   7   4   5
#175: 7   4   8   3
#176: 3   12   10   15

#177: 9	8	4	3
#178: 13	6	4	3
#179: 2	4	6	12
#180: 54	18	6	6
#181: 2	10	5	8
#182: 180	60	10	20
#183: 30	15	5	10
#184: 14	4	6	5
#185: 44	12	8	24
#186: 6	7	2	1
#187: 31	6	3	14
#188: 4	4	2	8
#189: 25	9	18	7
#190: 6	10	3	2
#191: 25	10	5	10
#192: 16	6	4	2
#193: 8	4	3	4
#194: 20	5	5	5
#195: 2	5	1	4
#196: 9	3	8	1
#197: 6	4	1	7
#198: 7	22	6	9
#199: 6	6	4	10
#200: 16	4	6	2
#201: 20	5	5	10
#202: 15	14	5	7
#203: 10	1	3	5
#204: 7	11	4	5
#205: 6	3	2	1
#206: 2	2	5	3
#207: 4	*	*	5	2

#208: 8	12	1	4
#209: 49	7	6	5
#210: 27	9	7	11
#211: 15	4	5	10
#212: 10	2	1	6
#213: 10	5	2	7
#214: 11	3	12	8
#215: 30	15	10	5
#216: 6	6	12	3
#217: 20	3	5	12
#218: 5	20	10	5
#219: 21	1	20	3
#220: 12	10	6	2
#221: 9	1	4	3
#222: 15	9	5	3
#223: 20	12	2	6
#224: 17	8	5	1
#225: 25	4	5	20
#226: 2	6	4	3
#227: 24	4	1	8
#228: 21	3	6	7
#229: 21	9	4	8
#230: 3	3	2	1
#231: 11	9	1	2	2
#232: 42	9	7	6
#233: 15	6	5	5
#234: 3	8	2	5
#235: 8	3	1	3	4
#236: 40	20	7	2	1
#237: 12	4	3	5
#238: 5	2	7	9

#239: 97   3   *   *   200
#240: 2   6   9   8
#241: 36   3   6   18
#242: 40   20   50   10
#243: 40   5   2   8
#244: 1   9   4   5
#245: 18   11   9   9
#246: 6   4   3   5
#247: 51   17   6   2
#248: 120   6   9   10
#249: 28   7   4   6
#250: 8   5   3   5
#251: 58   23   29   6
#252: 16   2   50   12
#253: 19   7   3   4
#254: 54   6   9   3
#255: 6   12   4   2
#256: 10   2   11   6
#257: 9   11   7   2
#258: 2   10   3   8
#259: 60   3   4   5
#260: 33   9   5   19
#261: 36   4   3   12
#262: 3   10   3   1
#263: 56   8   6   7
#264: 48   15   2   8
#265: 210   10   20   10
#266: 39   14   1   24
#267: 22   14   4   4
#268: 3   7   5   4
#269: 9   5   1   4   7

#270: 9   7   6   1
#271: 7   5   1   5
#272: 30   10   10   5
#273: 5   7   2   12
#274: 3   18   15   3
#275: 10   3   5   5
#276: 49   15   60   11
#277: 28   4   9   7
#278: 20   4   4   12
#279: 26   5   8   13
#280: 2   1   55   1   53
#281: 26   1   3   23
#282: 100   4   50   6
#283: 3   7   2   7   1
#284: 35   3   8   24
#285: 35   5   7   6
#286: 64   6   2   8
#287: 8   10   5   2
#288: 12   8   2   10
#289: 8   4   2   2   6
#290: 18   4   1   2   7
#291: 6   4   3   1   6
#292: 3   8   5   3   1
#293: 4   5   5   7
#294: 18   3   10   15
#295: 18   7   4   6
#296: 15   60   5   40
#297: 28   *   *   7   *
#298: 25   25   28   22
#299: 22   9   6   11
#300: 18   21   2   18

Made in the USA
Coppell, TX
15 June 2022

78853925R00063